TIME SPACE YOU AND

ORIGINS OF LIFE, LOVE, SEX AND GOD

AZAR

INDIA · SINGAPORE · MALAYSIA

Notion Press

No.8, 3rd Cross Street,
CIT Colony, Mylapore,
Chennai, Tamil Nadu – 600004

First Published by Notion Press 2020
Copyright © Azar 2020
All Rights Reserved.

ISBN 978-1-63633-774-6

This book has been published with all efforts taken to make the material error-free after the consent of the author. However, the author and the publisher do not assume and hereby disclaim any liability to any party for any loss, damage, or disruption caused by errors or omissions, whether such errors or omissions result from negligence, accident, or any other cause.

While every effort has been made to avoid any mistake or omission, this publication is being sold on the condition and understanding that neither the author nor the publishers or printers would be liable in any manner to any person by reason of any mistake or omission in this publication or for any action taken or omitted to be taken or advice rendered or accepted on the basis of this work. For any defect in printing or binding the publishers will be liable only to replace the defective copy by another copy of this work then available.

Contents

Introduction   5

   I. A Justification   7
  II. Physics and Metaphysics   9
 III. What Are We?   20
 IV. Concept of Time   31
  V. Where Are We?   41
 VI. The Elemental Question   48
 VII. Origin of Life   56
VIII. Unity of Lives   71
 IX. Origin of Sex   84
  X. Origin of Love   90
 XI. Romantic Love   99
 XII. Different Types of Love   109
XIII. Mature Love   118
XIV. Unity of Being   124

XV.	The Homosapien – A Natural History	128
XVI.	Unity of Doom	138
XVII.	Can God Save Us?	155
XVIII.	Disunity of Man	164
XIX.	Realizing Unity	173

References and Further Reading   *181*

Picture Credits   *185*

About the Author   *189*

Introduction

Making sense of the world around us and understanding what we are is a burning desire every human being experiences. While it may be dormant in some, it is overwhelming in others, while quite a good number are in between.

Though there are several myths, folklores and beliefs regarding the nature of our existence handed over to us over thousands of years, we are at a loss to know what to believe and what not to believe.

In this work we have taken the assistance of science, by way of facts as well as its methodology, in examining the nature of ourselves and this universe. Science is impersonal and cold and does not respect any statement which is not substantiated in a proper manner. This approach is to get as objective a view as possible of the ideas at stake.

The nature of Universe is examined from the day it originated and the human nature is analyzed from the day homosapiens evolved from the hominids to the present in this work. Pre-human and human origins of sex, love and community life has been dealt with in detail and it is shown how they had been instrumental in evolving the super consciousness of the humans.

Science is basically a study of nature – nature in its macro as well as the ultra-microscopic form. Any person who tries to understand nature more and goes deeper and deeper into it, be it a scientist or otherwise, gets dazed by the wonders which unfold before him. The beauty, harmony, unity and the unbelievable manifestations in which the universe presents itself goes beyond any human imagination. At this point, anyone becomes spiritual.

There is no quarrel between the spiritualists and scientists at this deepest level. There is only unity and harmony.

The pity is, bulk of our modern humans are at a loss to appreciate this unity and are thrust into a quagmire of anxiety, worries and apprehensions. The only way out is for them to realize this unity and their true self.

This work is for those who want to understand this world, realize what they are and enjoy the bliss and peace such awareness brings.

I. A Justification

A human being is a phenomenon in time and space.

At what time this phenomenon arrives or departs, how it happens or what is its destiny, we hardly have any dependable clues. Of course, we have too many speculations and assertions regarding them – but very little concrete ideas or, as technical people call it 'actionable data.'

To understand this phenomenon, we should also understand the context in which it happens – that is time and space, without which our understanding will be incomplete. If we do that, again we are led in to a quagmire. What appeared to be simple propositions become more and more mysterious and complex as we delve deeper into them. Einstein tore into the secrets of space and time and brought out facts and principles which govern our cosmos. After his explanations, this world looks more bewildering and enigmatic. Scientists of today agree with him and say space and time can be stretched, compressed or bent. For modern men like you and me, stretched by our worldly needs and bent on making a living on compressed resources, this looks like too big a boulder to break our headon.

In such a complex world system where do we stand?

Can the scientific knowledge available today give us answers to the umpteen questions human beings had been raising through the ages?

Can it give us an idea as to what we are, where we come from and where are we going?

This book is a humble attempt to address these issues and give a possible clarity on the phenomenon we are.

II. Physics and Metaphysics

"By nature, all men long to know."

– Aristotle, Metaphysics

The questions of what we are, where do we come from, what is our place in this world and what is our destiny have dogged the minds of our people from time immemorial. For some it is an existential question dominating their lives. But for a vast majority it is a dormant question, drowned in pursuits of worldly life, so that they do not raise their ugly heads. Nevertheless, everybody is uncomfortable with these questions and wish they had the answers. They keep trying to find out what they are and what the nature of this world is.

Twenty first century is an age of science. People need a crediblescientific explanation for anything and everything. They shred to pieces any formulation which does not stand to the strictest of such scrutiny possible.

We see images from Africa or Antarctica streamed live into our bedroom in real time. We know how it is done and why it is done. The same way, we know how the umpteen number of 'wonder' gadgets around us work or at least we believe the experts who designed them know.

Our scientific mind does not stop with this. We want to apply this scientific mind to entities beyond our physical world.

Physics and Metaphysics

Some of the most valuable things in human life are not physical – like your love for your family, your likes, dislikes, your ideas about life and death, your social responsibilities – the list is eternal. We try to make sense of the world with the available knowledge of the world around us.

But deeper and deeper we go into the nature of the physical world, we always come to a point where we understand that we have understood very little. The moment we arrive at an answer after elaborate scientific enquiry, we realize that the answers have generated more questions than the ones they answer. Then the speculations start. With available so called 'proven' data, we try to place the puzzles in perspective and try to understand some more of our world. In spite of putting all our known and speculative knowledge together, they shed very little light on the human needs beyond the realm of physical world. The needs of people to be happy, meaningful, satisfied and at peace with themselves and the world around them mostly go unaddressed.

Plato, the Greek philosopher who lived 2400 years ago, called these concerns beyond the visible and immediate physical world as 'Metaphysics' or 'Science beyond physics.' As per Greek wise men, Metaphysics is the branch of philosophy concerned with the nature of existence, being and the world. Arguably, metaphysics is the foundation of philosophy: Plato's student Aristotle calls it "first philosophy" (or sometimes just "wisdom"), and says it is the subject that deals with "first causes and the principles of things."

These Greek pondering happened long ago. Though we have progressed much over the 'physical' front, we do not seem to have moved much forward regarding the metaphysical part.

There is a large gap between the scientific discipline and the metaphysical realm of our time. The scientific inquiry often fails the metaphysical pursuits of our time, and ends up facing a high wall of uncertainty and ambiguity. People wonder what the purpose of life is and are perplexed whether they are lost in a desert of sand dunes or they are on track for an oasis.

This work aims to appeal to the modern men, tuned to the spirit of scientific inquiry and who seek reasonable answers to their metaphysical questions.

Our quest here is to see whether the abundant scientific physical knowledge of today can enable us to find answers to the 'metaphysical' questions faced by the confused humanity of the day. We will also strive to find out whether this knowledge can empower us to a little more wholesome, happier and satisfying life to our stressed, depersonalized fellow beings.

Metaphysics Started the Day Man Started Thinking

The philosophic questions about existence and the nature of the world we live in can be traced to the origin of human beings.

Charles Darwin, the man behind the evolutionary theories of the present, says in hisbook 'The Descent of Man' – 'As soon as the important faculties of the imagination, wonder, and curiosity, together with some power of reasoning, had become partially developed, man would naturally crave to understand what was passing around him, and would have vaguely speculated on his own existence.'

Man has always been preoccupied, overwhelmed or threatened by forces of nature which were beyond their understanding or power. Evidence for these can be found all over the world and in all cultures. We have found inscriptions in caves and rocks dated to pre historic times which denote 'super natural' entities or acts of their appeasement. Several artifacts retrieved from periods before the dawn of civilization indicate occult practices or placation of divinities was widely prevalent.

Deliberate, intentional burial, with rituals and accompanied by articles for 'after life'may signify a concern for the dead that transcends day to day life. It is so ancient, even Neanderthals, a separate species from us are supposed to have practiced such burial about 100,000 years back. Archaeological evidences also suggest that they might have had a form of animal worship.

Physics and Metaphysics ♦ **13**

The Löwenmensch figurine or Lion-man of the Hohlenstein-Stadel is a prehistoric ivory sculpture carved out of mammoth ivory using a flint stone knife. It was discovered in the Hohlenstein-Stadel, a German cave in 1939. The lion-headed figurine is the oldest-known zoomorphic (animal-shaped) sculpture in the world. It has been determined by carbon datingto be between 35,000 and 40,000 years old, associated with European Early Modern Humans.

1. **The Löwenmensch figurine.**

This lion-headed figurine is the oldest-known zoomorphic (animal-shaped) sculpture in the world. It has been determined by carbon dating to be between 35,000 and 40,000 years old, associated with European Early Modern Humans. The Löwenmensch figurine or Lion-man is a prehistoric sculpture carved out of mammoth ivory using a flint stone knife. It was discovered in the Hohlenstein-Stadel, a German cave in 1939.

2. Sorcerer, Grotte de Gabillou, France.

This is a horned and bearded figure. It belongs to the Upper Paleolithic and Mesolithic cultures in western Europe. Dated to be from around 17,000 to 12,000 years ago.

3. Bust of a goddess, perhaps Bau, wearing horned cap.

Limestone, Neo-Sumerian period (2150–2100 BC). From Telloh, ancient Girsu.

Physics and Metaphysics ♦ **15**

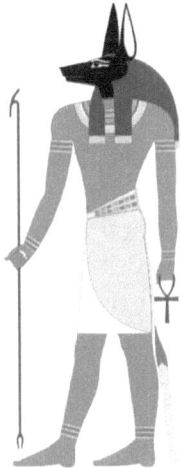

4. Anubis, the jackal headed god of ancient Egypt, was the god of embalming and the dead.

Since jackals were often seen in burial sites, the ancient Egyptians believed that Anubis watched over the dead.

5. Priest king of Mohenjo Daro – An iconic artifact of Indus valley civilization. It is dated 2200–1900 BC and was found at the Mohenjo Daro archaeological site, Sindh Province, Pakistan. Picture of a model kept at the Chennai Museum, India.

of animals. It also shows a man with a bird head and a bison. The Gabillou cave, a cave located in the French town of Sourzac shows a painting of a half-animal half-human being. It has been named 'The Sorcerer of Gabilou' and the archeologists believe it may be evidence of early super natural shamanic practices. Shamanic practices is the term used when a person who is supposed to have super natural powers and is able to interact with spirits, and can use this spirit power to bring about healing or some other material effect.

These Old Stone Age paintings have been assigned to the Cro-Magnons, the first members of *Homo sapiens* who settled in Europe. The paintings are attributed to the Magdalenian time which stands for some 11,000 to 17,000 years ago. Association of super natural and occult practices with a mixture of such human and animal figures is a subject which has been a topic of discussion for a long time.

The two most ancient civilizations of the world are that of the Sumerians and the Egyptians. They are more than five thousand years old. Both of them had great number of gods and elaborate rituals of worship. Their religious practices have been extensively documented in volumes of present day chronicles, showing how preoccupied they were with the 'other worldly' things.

Metaphysics for the Present

We still have several kinds of people who face these 'metaphysical' questions in their own way. First there are the 'believers,' who believe whatever has been handed over to them down the ages, by way of scripture, religious beliefs, wise sayings, local traditions etc. They feel utterly comfortable with the knowledge at their disposal. They feel any additional knowledge is not necessary or any new information is subservient to what they already know.

Then there are the 'indifferent,' who feel these metaphysical questions are not for them. They are happy with the day to day living and feel comfortable in not breaking their head over something on which a lot of people have already broken their heads and are still at a loss.

The third type of people are those who think such metaphysical questions are important, but find no rationale or scientific basis for the several questions raised like –what are we, where do we come from and where do we go, do we have a purpose or a destiny etc. etc. Not able to go any further, those questions are put aside under 'sleep mode.'

The 'believers' and the 'indifferent' also do get such questions once in a while, but push them out of their domain, either consciously or unconsciously. They also try to look into rationality of the beliefs they have.

Often, these questions attain enormous proportions and occupy one's mind, especially in times of great stress and difficult life situations. They may also be just vexed with the apparent meaninglessness of the endeavors they undertake and are unable to decide the direction of their life.

Whether a person likes it or not, questions concerning things beyond one's physical world are an integral part of one's psyche. Certain clarity and directions regarding this can definitely make a person more composed and happier.

People of present day have become too busy to contemplate on anything. This too much preoccupation also helps them avoid questions which can disturb them mentally. Either they engage in work or the so called 'pleasure' activities. However 'pleasurable' the activities are, ultimately they tend to become insipid, after the novelty fades and over exposure tires them. The core theme is to be engaged, and not to give vent to worrisome questions. Every time they get a break from this preoccupation, boredom, which seems to be the basic characteristic of humans, arises. With it arise uncomfortable questions and disquiet. This makes them mechanical and prone for anxieties. It prevents them from locking eyes with life and relishing its depth and beauty.

First we will see what scientific method is. We can roughly describe it, without getting into too many technical nuances as

'A method or procedure characterized by systematic observation, measurement, and experiment, and the formulation, testing, and modification of hypotheses.'

We can also add that criticism is the backbone of the scientific method and any time it can be challenged or changed in the light newer facts as they make their appearance.

With reference to the above description, we will go entirely by proven and well documented facts of the scientific world as of today. We will not go by any fact which has not been documented, tested or of doubtful origin.

The questions we will try to answer are the eternal ones like

What are we?

Where are we?

Why are we here?

Do we have purpose?

Do we have a destiny?

How do we possibly end? And

Is there a way forward?

III. What Are We?

"Physicists are made of atoms. A physicist is an attempt by an atom to understand itself."

– Michio Kaku

If we really want to find out what we are and where we come from, we have to dig into our past, as past is a repository of knowledge and only source from where we can make some predictions. The present is fleeting and future is a mystery. So, we will get back to the farthest time possible, backwards and examine it in the light of all available and relevant knowledge of today to pick up some clues.

The farthest time we can go back at our present state of scientific knowledge is the 'Big Bang,' during which everything, includingthe space and time is supposed to have originated.

So, let us not make any compromise and start our inquiry with a consideration of the Big Bang itself.

Everybody is not convinced about the occurrence of Big Bang and there is much controversy surrounding it. But scientists and astronomers from varied countries and varied scientific disciplines are now convinced that the 'Big Bang' happened and is the best explanation possible for cosmic phenomena as we see them now. Scientists have also produced irrefutable evidences for the same and the theory has withstood much criticism and investigation.

In the most common models, initially the universe was filled homogeneously and iso tropically with an incredibly high energy density and huge temperatures and pressures. You may be surprised by the term 'High energy density,' as at that time both energy and density was one and was beyond any measurement by physical methods of today.

6. The Big Bang was so cataclysmic that it left scars on the fabric of the cosmos that we can still detect today: A picture by NASA.

Suddenly, this unity called high energy density started rapidly getting inflated. This happened about 13.8 billion years ago. As it rapidly expanded in fraction of a millisecond, the universe grew exponentially. After what is called 'cosmic inflation' stopped, the universe consisted of quark–gluon plasma, as well as all other elementary particles.

Before following the events of the Big Bang, it will be necessary for us to go little deeper in to the micro particles which make up our universe.

People from ancient times had been wondering what our physical world is made of. It was Greek philosopher Leucippus who lived in 5th century BCE, along with his disciple Democritus, developed the idea that everything is composed entirely of various imperishable, indivisible elements called atoms. The word 'a-tom' in Greek meant 'un-cuttable' or a smallest unit that cannot be divided further. They are still remembered today for their formulation of an atomic theory of the universe. But the theory today seems to be bursting in to fragments as our scientists have managed to cut and divide the atom. Now our atomic labs are discovering such small fragments of the atom that they have started wondering whether it is a particle or an energy oscillation.

Coming back to the Big Bang, a quark is a type of elementary particle and a fundamental constituent of matter. Quarks combine to form composite particles called hadrons, the most stable of which are protons and neutrons, the components of atomic nuclei. Quarks are never directly observed or found in isolation; they can be found only within hadrons.

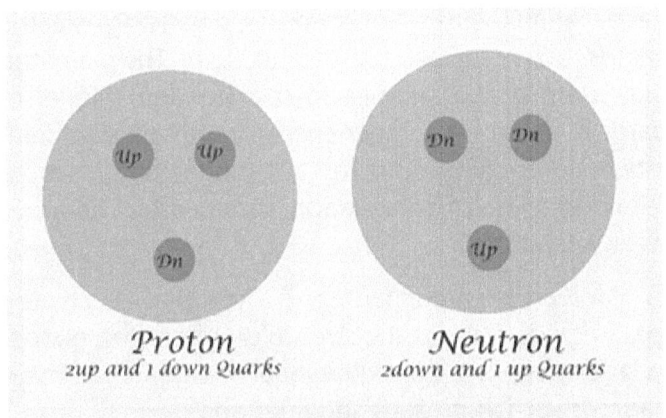

7A. Quarks, Proton and Neutron.

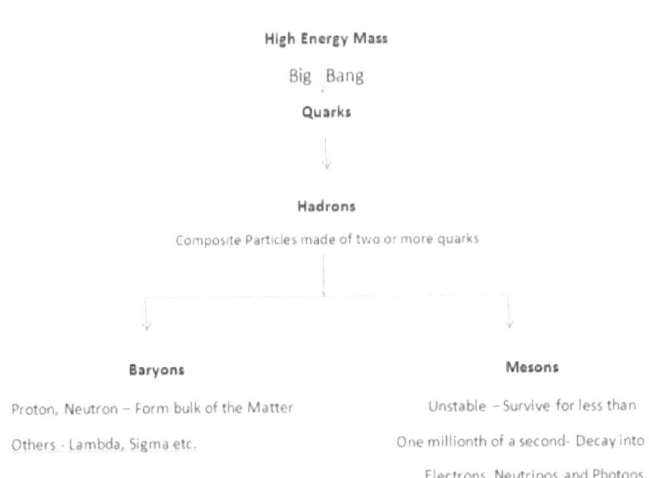

7 B. Big Bang and formation of matter.

Quarks are the only known particles whose electric charges are not integer multiples of the elementary charge, which means they are single units. Quarks have various intrinsic properties, including electric charge, mass, color charge and spin. They are the only elementary particles in the Standard Model of particle physics to experience all four fundamental interactions, also known as fundamental forces (electromagnetism, gravitation, strong interaction, and weak interaction).

A gluon is an elementary particle that acts as the exchange particle for the strong force between quarks. In layman's terms, they "glue" quarks together, forming hadrons such as protons and neutrons.

In particle physics, a hard on is a composite particle made of two or more quarks held together by the strong force in a similar way as molecules are held together by the electromagnetic force. Hadrons are categorized into two families – baryons and mesons.

Baryons are made of an odd number of quarks – usually three quarks. Protons and neutrons are examples of baryon sand most of the mass of ordinary matter comes from two hadrons, the proton and the neutron.

Mesons are sub-atomic particles made up of a combination of quarks and anti quarks. Mesons are made of an even number of quarks—usually one quark and one anti quark. They can annihilate each other once they come near each other and so exist for a very short time, for less than a millionth of a second.

Baryogenesis – Formation of Matter

After inflation stopped, temperatures were so high that the random motions of particles were at unimaginable speeds, and particle – antiparticle pairs of all kinds were being continuously created and destroyed in collisions. At some point, an unknown reaction called baryogenesis violated the conservation of baryon number, leading to a very small excess of quarks and leptons (leptons are subatomic particles which can be charged as in electrons or neutral as in neutrinos) over anti quarks and antile ptons—of the order of one part in 30 million. This resulted in the predominance of matter over antimatter in the present universe

In physical cosmology, baryogenesis is the hypothetical physical process that took place during the early universe that produced baryonic asymmetry, i.e. the imbalance of matter (baryons) and antimatter (antibaryons) in the observed universe. If we are to talk without the nuclear jargon, it simply means that when matter and antimatter were in equal measures, no matter could be formed. When there was a small excess of matter over antimatter, matter was formed, which is called baryogenesis.

Baryogenesis was followed by primordial nucleosynthesis, when atomic nuclei began to form. The components of nuclei are basically protons and neutrons. Two up quarks and one down quark formed a proton while one up quark and two down quarks formed a neutron. Most elements are formed either with only proton in their nucleus (as in hydrogen atom) or a combination of several protons and neutrons in the nucleus, as in heavier elements.

In physical cosmology, Big Bang nucleosynthesis, also known as primordial nucleosynthesis, refers to the production of nuclei other than those of the lightest isotope of hydrogen during the early phases of the Universe. Primordial nucleosynthesis is believed by most cosmologists to have taken place in the interval from roughly 10 seconds to 20 minutes after the Big Bang, and is calculated to be responsible for the formation of most of the universe's helium as the isotopes of helium and hydrogen.

While lighter elements formed straight away, heavier elements had to take a circuitous route, which we will see later.

This is the matter, which originated from a unitary high energy mass and gave birth to everything in this earth – sun, moon, stars, earth, air, water, the food we eat, you, me, the people we live with or fight with.

All people do not agree with the nature of fundamental particles as discussed above and they have alternative theories to put forward.

String Theory and Superstring Theory

In the world of theoretical physics, the latest theory which attempts to unify all theories regarding the nature of universe is called the 'string theory.' This theory questions the very concept of these subatomic particles. This theory too has experimental evidence and solid backing from several scientists.

The original version of string theory, developed in the late 1960s is known as Bosonic string theory. It is called so because it contains only bosons in the spectrum.

The starting point for string theory is the idea that the point-like particles of particle physics can also be modeled as one-dimensional objects called strings. The string theory states that there is only one kind of string, which may look like a small loop or segment of ordinary string and it can vibrate in different ways.

Each string has a particular vibration mode (i.e. it vibrates in a particular way). The number of times a string vibrated in a unit time (frequency) corresponds to each particle and thereby determines the mass and size of the particle. Simply put, the way a string vibrates actually determines the mass and size of a particle.

Each mode of vibration represents a distinct particle. So, these strings, by their different vibrations, make different types of matter. They also make gravity, electromagnetism, the strong force and the weak force. These strings are very small, measuring about a millionth of a billionth of a billionth of a millimeter and impossible to be seen with present technology.

The importance of string theory is that it can explain the nature of both matter and space-time. String theory also answers a series of perplexing questions about the bewildering number of subatomic particles, such as why there are so many of them in nature.

To understand the string theory, you can imagine the strings secured under tension on a guitar. When strung, the guitar strings produce different types of sound, depending on the type of vibration induced. Since there are an infinite number of harmonies that can be composed for the guitar, there are an infinite number of forms of matter that can be constructed out of vibrating strings. The universe itself, composed of countless vibrating strings, would then be comparable to a symphony.

These strings oscillate in millions of different ways, and create the elemental particles which, in turn, create atoms. These atoms create molecules which bond together to make water, sand, trees, animals or anything in the universe.

In the 1980s, supersymmetry was discovered in the context of string theory. Supersymmetry is a principle that proposes a relationship between two basic classes of elementary particles: bosons, and fermions

There were five separate superstring theories which caused a great confusion among the physicists. Since the beginning of what is called the second superstring revolution in the 1990s, the five superstring theories are regarded as different limits of a single theory. According to the theory, the fundamental constituents of reality are strings of the Planck length (about 10–33 cm) that vibrate at resonant frequencies. Every string, in theory, has a unique harmonic. Different harmonics determine different fundamental particles. The graviton (the proposed messenger particle of the gravitational force), for example, is predicted by the theory to be a string with wave amplitude zero.

It is also suggested that the five string theories might be different limits of a single underlying theory, called M-theory.

What does the M stand for?

The scientists are yet to determine what it is!

Eminent physicist and protagonist of superstring theory Michio Kaku states, "What is Universe? The Universe is a symphony of vibrating strings… We are nothing but melodies. We are nothing but Cosmic Music played out on vibrating Strings and Membranes"

Superstring theory gives us a compelling formulation of the theory of the universe, and offers explanation for several phenomena the conventional physics could not explain. But a fundamental problem is that experimental tests of the theory seem beyond our present-day technology. Another notable feature of string theories is that these theories require extra dimensions of space and time for their mathematical consistency, whose details we will see later

As for now, we end up with the question whether we are made up of 'particles' or 'strings.'

But whatever our fundamental nature is, there is no doubt we originated from the same 'unitary high energy mass' from which the whole universe originated.

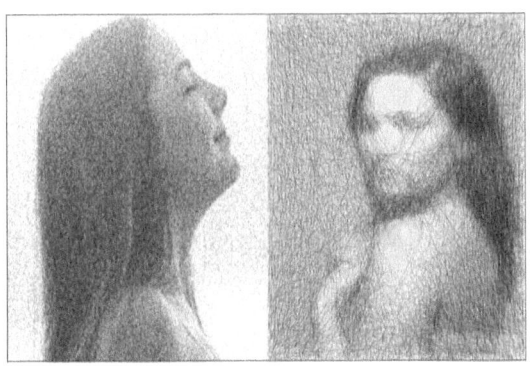

**8. What are you?
Particles or Strings!**

Another very intriguing fact is that top cosmologists around the world believe that time and space also originated with the 'Big Bang.'

So, we will have to look into the concept of time and space, the integral aspects of the cosmos, if we are to go any further as to discover 'What are we?'

IV. Concept of Time

'People like us, who believe in physics, know that the distinction between past, present and future is only a stubbornly persistent illusion'

– Einstein

We assume the concept of time is self-evident. An hour consists of a certain number of minutes, a day consists of hours and ayear consists of days. But we rarely think about the fundamental nature of time.

The early Christian theologian St. Augustine (4th–5th Century CE) said it aptly – "What then is time? If no one asks me, I know; if I wish to explain it to one that asks, I know not." He was only able to conclude that time was some kind of a "distention" of the mind which allows us to simultaneously grasp the past in memory, the present by attention, and the future by expectation.

In Greek mythology, Khronos was the personification of time. The figure of Khronos was typically portrayed as a wise old man with a long grey beard. The Horae or Hours were the goddesses of the seasons and the natural flow of time, generally portrayed as personifications of nature in its different seasonal aspects. The cycle of the seasons themselves were symbolically described as the dance of the Horae. At least we understand now how the word 'hour' originated.

To digress alittle, most people wonder how the hour was divided into an inconvenient number of 60 minutes. The division of the hour into 60 minutes and of the minute into 60 seconds comes from the Babylonians who used a sexagesimal (counting in 60s) system for mathematics and astronomy. We can see such divisions in longitudes and latitudes and divisions of a circle also.

Ancient Ideas

The early Greek philosophers generally believed that the universe (and therefore time itself) was infinite with no beginning and no end. In the 5th Century BCE, the Sophist philosopher Antiphon asserted that time is not a reality (*hypostasis*), but a concept (*noêma*) or a measure (*metron*). Also in the 5th Century BCE, Parmenides believed that reality was limited to what exists in the here and now, and the past and future are unreal and imaginary. His near-contemporary Heraclitus, on the other hand, firmly believed that the flow of time is real and the very essence of reality.

Christianity and the other Abrahamic faiths, Islam and Judaism, believed in an all-powerful and infinite God. As a logical extension medieval Christian, Muslim and Jewish philosophers and theologians developed the concept of the universe having a finite past with a definite beginning (the moment of its creation by God) and a definite end. Time, therefore, was necessarily finite in nature, a doctrine known as 'Temporal Finitism.' The general Christian view is that time will come to a definite end with the end of the world, in the so-called "end-times." This concept of impending doomsday, when the world could come to a final end could be seen in many religions around the world.

There are also extensive references in Hindu scripture about 'Kaliyuga,' a period when world will come to an end, but possibly to restart again.

Wheel of Time

In ancient Indian philosophy, as expounded in early texts such as the *Vedas* of the late 2nd millennium BCE, the universe goes through repeated cycles of creation, destruction and rebirth (with each cycle lasting 4,320 million years according to some sources). This led to a cyclic view of time, the so-called "wheel of time" or Kalachakra, in which there are repeating ages over the infinite life of the universe. This was coupled with a belief in an endlessly repeated cycle of rebirths and reincarnations for individuals. The wheel of time concept is found in Hinduism and Buddhism, as well as in the beliefs of the ancient Orphics and Pythagoreans of Greece, the Mayans of Mexico, the Q'ero Indians of Peru and the Hopi Indians of Arizona.

9. Kal Chakra – Wheel of Time, as per Hindu Mythology.

The idea of time as consisting of endlessly repeated cycles is perhaps an unsurprising one given the observed repetitiveness of other natural phenomena, such as the day-and-night cycle, the motion of the tides, the monthly cycle of the Moon, the annual cycle of the seasons, etc.

Modern Ideas of Time

Modern 'Global' people as we are, we do have some practical problems with time. If we talk to a friend from one side of the globe to the other, we feel the difference. If we say it is 10 o'clock in the morning, the other person says it is 10 at night in his place.

The same way, people in Tokyo, Addis Ababa, Berlin or Mexico will quote their own time. Of course, this variation in time and its geographical reasons are highly understandable to everyone.

Imagine yourself watching the star Proxima Centauri, which is the closest star to earth, if we exclude the sun. It is four light years away from us, meaning it takes four years for its light to reach us. If there had been any activity or explosion of this star, it will take four years for us to know. To put it another way, what we are seeing of this star right now is how it was four years ago. How it is or what it is right now at the present moment, we have absolutely no knowledge.

Similarly the Andromeda Galaxy also known as Messier 31, which is the nearest major galaxy to our own Milky Way galaxy is about 2.5 million light-years from Earth.

MACS1149-JD1 is one of the farthest known galaxies from Earth and is about 13.28 billion light years away from us. In case a star explodes in any of these galaxies, we will take millions or billions of years to know it. We are not even sure mankind will survive that long to witness the event. What we are doing now is observing the celestial bodies as they existed millions or billions of years ago and summing them up as the present moment for us. What is their status right now or how many billions of kilometers they have moved from our present observed position, it is difficult even to imagine.

Between those galaxies and us, we do not know what the time is or how it is progressing.

Suppose there is a celestial person out there in the cosmos and asks another such person "What is the time now?" it is very likely the answer will be "In reference to which galaxy or to which 'event' or on what scale."

Newtonian Time

The most famous proponent of Absolute Time is Sir Isaac Newton, which is also sometimes known as "Newtonian time" in his honour. He says time exists independently of any perceiver, progresses at a consistent pace throughout the universe, is measurable but imperceptible, and can only be truly understood mathematically. For Newton, absolute time and space were independent and separate aspects of objective reality, and not dependent on physical events or on each other.

But more modern concepts, after the advent of Relativity Theories of Einstein, have taken the concept of time to unimaginable complexities.

Relativistic Time

According to the general theory of relativity, space, or the universe, emerged in the Big Bang some 13.7 billion years ago. Before the Big Bang, there was no space or time.

"In the theory of relativity, the concept of time begins with the Big Bang the same way as parallels of latitude begin at the North Pole. You cannot go further north than the North Pole," says Kari Enqvist, Professor of Cosmology.

One of the most peculiar qualities of time is the fact that it is measured by motion and it also becomes evident through motion.

Since Albert Einstein published his Theory of Relativity (the Special Theory in 1905, and the General Theory in 1916), our understanding of time has changed dramatically. The traditional Newtonian idea of absolute time and space has been superseded by the notion of time as one dimension of space-time in special relativity, and of dynamically curved space-time in general relativity.

It was Einstein's genius to realize that the speed of light is absolute, invariable and cannot be exceeded. In relativity, time is certainly an integral part of the very fabric of the universe and cannot exist apart from the universe. Einstein realized that if the speed of light is invariable and absolute, both space and time must be flexible and relative to accommodate this.

Much of Einstein's work is often considered "difficult" or "counter-intuitive." But his theories have been proved to be a remarkably accurate model of realityboth in laboratory experiments and in astronomical observations. They are much more accurate than Newtonian physics, and applicable in a much wider range of circumstances and conditions.

Space-Time

One aspect of Einstein's Special Theory of Relativity is that we now understand that space and time are merged inextricably into four-dimensional space-time. With this insight, time effectively becomes just part of a co-ordinate specifying an object's position in space-time.

Quanta or 'Packets Theory' and Time

In the first half of the twentieth Century, a whole new theory of physics was developed. This theory has superseded everything we know about classical physics, including the Theory of Relativity, which is still a classical model at heart. Quantum theory or quantum mechanics is now recognized as the most correct and accurate model of the universe, particularly at sub-atomic scales, although for large objects classical Newtonian and relativistic physics work adequately.

One of the implications of quantum mechanics is that certain aspects and properties of the universe are quantized, i.e. they are composed of discrete, indivisible packets or quanta. An obvious question, then, would be: is time divided up into discrete quanta? According to quantum mechanics, the answer appears to be "no," and time appears to be in fact smooth and continuous. However, if time actually is quantized, it is likely to be at the level of Planck time (about 10^{-43} seconds), the smallest possible length of time according to theoretical physics, and probably forever beyond our practical measurement abilities.

Bending of space-time

With the General Theory of Relativity, Einstein has formulated and proved that in regions of very large masses, such as huge stars and black holes, space-time is bent or warped substantially by the extreme gravity of the masses.

Gravitational Time Dilation

Similarly, time dilatation has also been demonstrated and scientists call it Gravitational Time Dilation. This effect measures the amount of time that has elapsed between two events by observers at different distances from a gravitational mass. They have observed that time runs slower wherever gravity is strongest, and this is because gravity curves space-time.

String Theory and Multi-Dimensional Space-time

String theory, the theory which attempts to unify all theories regarding the nature of universe, has its own take on time and space.

A notable feature of string theories is that these theories require extra dimensions of spacetime for their mathematical consistency. In bosonic string theory, spacetime is 26-dimensional, while in superstring theory it is 10-dimensional, and in M-theory it is 11-dimensional.

They also predict the possibility that parallel universes, apart from the one we are living in, might exist with their own space and time.

It is pretty obvious now that as we try to go deeper into the understanding of time, more confusion we encounter.

We again get reminded of Einstein's words: 'People like us, who believe in physics, know that the distinction between past, present and future is only a stubbornly persistent illusion.'

So, what is time?

May be we need some more time to resolve this question!

V. Where Are We?

Is there only one space for us to live in?

When, after a long day of running around, you finally find the time to relax in your favorite armchair, nothing seems easier than just sitting still. You think you are static like a stone.

But have you ever considered how fast you are really moving when it seems you are not moving at all?

For the earth to make one complete rotation in 24 hours, a point near the equator must move at close to 1600 km/hour. The speed gets less as you move north or south towards the poles, but still it moves. Because gravity holds us tight to the surface of our planet, we move with the earth and do not notice its rotation in everyday life.

In addition to spinning on its axis, the Earth also revolves around the Sun. We are approximately 150 million km from the Sun, and at that distance, it takes us one year (365 days) to go around once. The full path of the Earth's orbit is close to 970 million km.

To go around this immense circle in one year we are travelling at a speed of 107,000 km/hr.

Next we can focus on the speed of the Sun around the center of the Milky Way Galaxy.

It takes our Sun approximately 225 million years to make the trip around our Galaxy, the Milky Way. This is sometimes called our "galactic year." How fast do we have to move to make it around the Milky Way in one galactic year? It's a huge circle, and the speed with which the Sun has to move is an astounding 792,000 km/hour! The Earth, anchored to the Sun by gravity, follows along at the same fantastic speed. Of course, perched on the earth, we also go on this cosmic ride at this break neck speed.

Not only this! The galaxies are also moving away from each other at tremendous speeds. Without getting into the intricacies of physics like reference points for the speed etc. (as speed is also relative), we can be sure we are travelling at a speed of hundreds of thousands of kilometers per hour, non-stop, day in and day out. In case an alien happens to be suspended in the cosmos nearby, and he wants to show his wife where you are, I am sure he will have great difficulty and get cynical disapprovals from her. His fingers will have to move all the time to keep pointing at us. That is our status – we do not even have a place to call our own and speed along the universe at millions of kilometers a day, perched on a tiny celestial body called the earth.

Add to this some complexities involved in space. Scientists say the expansion of the universe is not like any other expansion. *When the universe expands, it is space itself that is stretching.* The galaxies in the universe are moving apart because space stretches and creates more distance between them. It is not something like your moving inside a room. It is like as you move more and more to the periphery of the room, the room itself is becoming bigger and bigger and it is endless.

All it means is, not only our position in space is constantly changing, but even the space itself is changing.

This seems to have some relevance to the Uncertainty principle, also called Heisenberg Uncertainty Principle or Indeterminacy principle, put forward by the German physicist Werner Heisenberg. He says that the position and the velocity of an object cannot both be measured exactly, at the same time, even in theory. The very concepts of exact position and exact velocity together, in fact, have no meaning in nature. As per this theory, even the quest to realize our place in the space looks meaningless.

Superstring Theory and Multi-Dimensional Space

Theodor Franz Eduard Kaluza, a German mathematician and physicist, published a research work in 1921, which proposed more dimensions to space time than the standard four.

In physics, Kaluza–Klein theory (KK theory) is a classical unified field theory of gravitation and electromagnetism. It is built around the idea of a fifth dimension beyond the usual four of space and time and considered an important precursor to string theory.

Many physicists, including several Nobel laureates, are now convinced that a conventional four-dimensional theory is inadequate to describe the forces that define our universe. They suggest the universe may actually exist in higher-dimensional space, called Hyperspace.

Scientifically, the Hyperspace theory goes by the names of Kaluza–Klein theory and supergravity. Its most advanced formulation called superstring theory, predicts the precise number of dimensions as ten, the usual three dimensions of space (length, width, and breadth) and one of time are now extended by six more spatial dimensions. The dimensions will be: 3D regular space + 1 time + 6D hyperspace. .

In bosonic string theory, spacetime is 26-dimensional, while in superstring theory it is 10-dimensional, and in M-theory it is 11-dimensional.

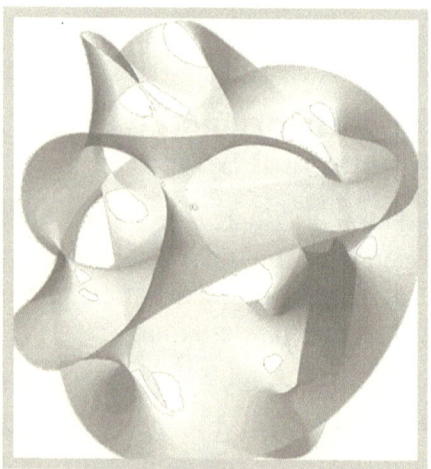

10. Computer depiction of complex multidimensional curves and spaces.

Then why is that we see only 3 dimensions of space?

Physicists suggest that the extra dimensions are restricted and have given technical explanations like compactification of extra dimensions to a very small scale, or that our world may live on a 3-dimensional submanifold corresponding to a brane, on which all known particles besides gravity would be restricted.

In the hyperspace theory, "matter" can be also viewed as the vibrations that ripple through the fabric of space and time. Thus follows the fascinating possibility that everything we see around us, from the trees and mountains to the stars themselves, are nothing but vibrations in hyperspace. If this is true, then this gives us an elegant, simple, and geometric means of providing a coherent and compelling description of the entire universe.

The Hyperspace physics have led the cosmologists to propose the startling possibility that our universe is just one among an infinite number of parallel universes. These universes might be compared to a vast collection of soap bubbles suspended in air. On each bubble, we can define our own distinctive space and time,

Central to this revolutionary perspective on the universe is the realization that higher-dimensional geometry may be the ultimate source of unity in the universe. Simply put, the matter in the universe and the forces that hold it together, which appear in a bewildering, infinite variety of complex forms, may be nothing but different vibrations of hyperspace.

However, the fact that our universe is curved in an unseen dimension beyond our spatial comprehension has been experimentally verified by a number of rigorous experiments. This gives rise to the possibility of existence of wormholes, which is the shortest route that can connect two such regions in different time – space zones.

If wormholes exist (a big IF), it gives several exciting possibilities. Wormholes may connect a universe with itself, perhaps providing a means of interstellar travel. Since wormholes may connect two different time eras, they may also provide a means for time travel. Wormholes may also connect an infinite series of parallel universes. The hope is that the hyperspace theory will be able to determine whether wormholes are physically possible or merely a mathematical curiosity.

Though all the above may sound like science fiction to some, the above theories and practical implications are rigorously worked on by some of the world renowned and ablest of scientists the world over. The world of scientific research is a never ending process of questions and answers and we will have to wait and see where it takes us.

Shall we take this to imply we cannot be sure where we are right now or at what speed we are going or where we are going?

The more we try to understand our place in time and space, more elusive they seem to be.

This again reminds us our previous notion that, the deeper and deeper we tend to analyze the nature of this physical world, more and more we are exposed to the limitations of our knowledge. Our need to jump from the 'physical' to the 'metaphysical' seems inevitable.

But a few very basic and elementary questions have to be understood before we go into some major clarifications.

VI. The Elemental Question

The chemicals you were made of where cooked in the hearts of exploding stars. Every bit of you is miraculous

– Brad Jenson

According to the Big Bang theory, the temperatures in the early universe were high and fusion reactions could take place. But only the lightest elements, hydrogen, deuterium, helium and small amounts of lithium and berylliumwere produced in the Big Bang nucleosynthesis.

As the universe expanded after the Big Bang, temperatures started coming down. The extremely high temperatures required for the creation of elements heavier than lithium or berylliumby nuclear reactions was simply not there. Lithium, with atomic number 3 and beryllium with atomic number 4 are light elements, which implies they have only 3 or 4 protons respectively in each one of their atoms. Most elements, including oxygen, carbon and nitrogen which form major component of our body and umpteen other elements which were necessary for our body and the formation of earth have much heavier atomic weight than the above two elements which formed during the big bang.

But unfortunately the universe was too cool to form these elements, even though it had the raw materials to form them.

Then how did these heavy elements which form an integral part of our universe and our human body form?

For this, we have to follow our wonderful cosmic story further.

There are about 90 naturally occurring elements which are present throughout the universe.

Two elements comprise around 98% of all the elements in the universe, with hydrogen at 75% and helium at 23%. These elements formed clouds of dust that were scattered throughout most galaxies, after the Big Bang to the present.

Some disturbance deep within these clouds gave rise to some focal points with mass such that the gas and dust begin to accumulate at those points. As they accumulate, they collapse under their own gravitational attraction. As the cloud collapses, the material at the center begins to heat up. Now, known as a protostar, it is this hot core at the heart of the collapsing cloud that will one day become a star. Not all of this material in a cloud ends up as part of a star—the remaining dust can become planets, asteroids, or comets or may remain as dust.

Thus a star is born.

Main Sequence Stars

Stars are fueled by the nuclear fusion of hydrogen to form helium deep in their interiors. The outflow of energy from the central regions of the star provides the pressure necessary to keep the star from collapsing under its own weight, and the energy by which it shines.

All the stars convert hydrogen into helium by nuclear fusion. In stars less massive than the Sun, this is the only reaction that takes place. In stars more massive than the Sun (but less massive than about 8 solar masses), further reactions that convert helium to carbon and oxygen take place. In the very massive stars, the reaction chain continues to produce elements like silicon up to iron.

Massive stars need higher central temperatures and pressures to support themselves against gravitational collapse, and for this reason, fusion reactions in these stars proceed at a faster rate than in lower mass stars. The result is that massive stars use up their core hydrogen fuel rapidly and spend less time on the main sequence before evolving into a red giant star.

The lifetimes of main sequence stars therefore range from a million years for a 40 solar mass O-type star, to 560 billion years for a 0.2 solar mass M-type star. Given that the Universe is only 13.7 billion years old, it means that small M-type stars that have ever been created are still in their prime. It also means that numerous massive stars have already exploded. The Sun, a G-type star with a main sequence lifetime of about 10 billion years, is currently 5 billion years old – about half way through its main sequence lifetime.

Modern science has outlined the specific nuclear reactions that occur in stars and supernovas to form the heavy elements. When a star's core runs out of hydrogen, the star begins to die out. The dying star expands into a red giant, and this now begins to manufacture carbon atoms by fusing helium atoms.

More massive stars begin a further series of nuclear burning or reaction stages. The elements formed in these stages range from oxygen through to iron.

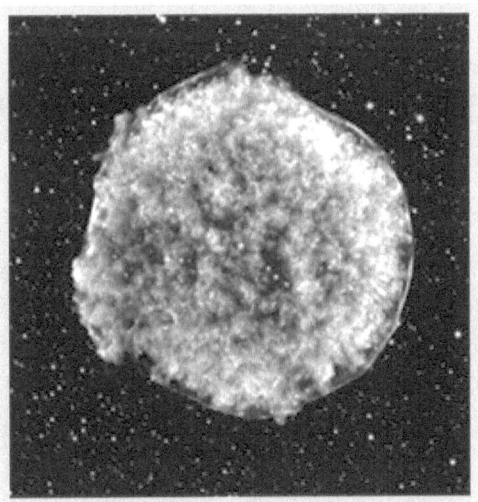

11. Tycho Supernova.

In 1572, Danish astronomerTycho Brahe noticed a new bright object in the constellation Cassiopeia. Tycho

showed that this "new star" was far beyond the Moon, and that it was possible for the universe beyond the Sun and planets to change.

Astronomers now know that Tycho's new star was not new at all. Rather it signaled the death of a star in a supernova, an explosion so bright, it outshone the light from an entire galaxy.

Nucleosynthesis, the process of formation of new elements, requires a high-speed collision, which can only be achieved with very high temperature. The minimum temperature required for the fusion of hydrogen is 5 million degrees. Elements with more protons in their nuclei require still higher temperatures. For instance, fusing carbon requires a temperature of about one billion degrees! Most of the heavy elements, from oxygen up through iron, are thought to be produced in stars that contain at least ten times as much matter as our Sun.

Can we manufacture them on Earth? The conditions and the unimaginable temperatures they require is impossible even to dream of on earth. At such high temperatures, the earth itself will disappear in seconds.

As the star runs out of nuclear fuel, some of its mass flows into its core. Eventually, the core is so heavy that it cannot withstand its own gravitational force. The core collapses, which results in the giant explosion of a supernova. During this supernova stage, the star releases very large amounts of energy as well as neutrons, which allows elements heavier than iron, such as uranium and gold, to be produced. In the supernova explosion, all of these elements are expelled out into space.

Our world is literally made up of elements formed deep within the cores of stars which had exploded long back. Their debris were made available in the solar nebula for the creation of our solar system and what we are.

The elements contained in your body are older than even the Sun and the solar system.

You can proudly claim the chemicals you were made of where cooked in the hearts of exploding stars.

The Birth of our Place-Nebular Hypothesis

According to this theory, the Sun and all the planets of our Solar System began as a giant cloud of molecular gas and dust. The giant cloud contained Hydrogen, Helium and some light elements which made their appearance during the Big Bang and heavier elements like Carbon, Oxygen, iron and other elements manufactured in the core of previously exploded massive stars.

Then, about 4.57 billion years ago, something happened that caused the cloud to collapse. This could have been the result of a passing star, or shock waves from a supernova, but the end result was a gravitational collapse at the center of the cloud.

From this collapse, pockets of dust and gas began to collect into denser regions. As the denser regions pulled in more and more matter due to their increasing gravitational pull, conservation of momentum caused it to begin rotating. This also increased the pressure inside and caused it to heat up. Most of the material ended up in a ball at the center while the rest of the matter flattened out into disk that circled around it. While the ball at the center formed the Sun, the rest of the material would form into the protoplanetary disc, that is a precursor of the planets.

12. **Solar nebula – The solar system in the making.**

The planets formed by the accumulation of particles into a massive object by gravitationally attracting more matter, a process called accretion. So, dust and gas gravitated together and coalesced to form ever larger bodies. Due to their higher boiling points, only metals and silicates could exist in solid form closer to the Sun, and these would eventually form the terrestrial planets of Mercury, Venus, Earth, and Mars. They had a solid outer surface because of their content. Because metallic elements only comprised a very small fraction of the solar nebula, the terrestrial planets could not grow very large.

In contrast, the giant planets (Jupiter, Saturn, Uranus, and Neptune) formed beyond the orbit of Mars, the 'Frost Line,' were cool enough for volatile icy compounds to remain solid. The ices that formed these planets were more plentiful than the metals and silicates that formed the solid inner planets. This allowed them to grow massive enough to capture large atmospheres of hydrogen and helium.

Of all the planets that were formed around the sun, one planet had unique characteristics. It had the right temperature, right atmosphere and right chemicals to form a wonder called 'Life.'

How this miracle came to be, itself is a fascinating story.

VII. Origin of Life

What was created first – The Soup or the Soul?

When earth started its journey as a celestial body going around the sun, there were no living organisms riding on it. The process by which life formed on earth is called 'abiogenesis,' which technically means formation of life from non-living or inanimate substances.

There are several theories explaining the appearance of life. Some of them worth considering and some, simply outrageous. There are people who postulate life has been brought to earth through rocks which regularly get blasted off Mars by cosmic impacts or somewhere else from the space. This concept is called 'Panspermia.' Again this begs the question of how life formed in the outer space or in the Mars.

There are also claims life could have formed under thick sheets of ice. Ice might have covered the oceans 3 billion years ago, as the sun was about a third less luminous than it is now. This layer of ice, possibly hundreds of feet thick, might have protected fragile organic compounds in the water below from ultraviolet light and destruction from cosmic impacts. This might have enabled these molecules to survive longer, allowing key reactions for the appearance of life. Again, this needs more substantiation.

Primordial Soup

The theory most commonly accepted by the authorities doing research on the origin of life is the Primordial Soup Theory. It states that life began in a warm pond or ocean from a combination of chemicals that could form amino acids, which then create proteins. This is supposed to have happened nearly four billion years ago. The process is also called 'Abiogenesis' as life is thought to have originated from nonliving substances.

The Russian Chemist A.I. Oparin and English Geneticist J.B.S. Haldane first conceived the 'Primordial Soup Theory' regarding the origin of life. Both developed this theory independently in 1920.

In this theory, the basic building blocks of life came from simple molecules which formed on the surface of the earth. The rain from the atmosphere created the "organic soup" with these molecules and it was then energized by lightning. It is possible the first organisms were consuming other organisms for energy and survival (heterotrophs) as photosynthesis had not made its appearance and they could not manufacture their own food.

Experimental support for the possible origin of life came in 1952. Chemist Stanley Miller and physicist Harold Urey did a remarkable experiment to test this theory. They mixed gases thought to be present on primitive earth: They were Methane (CH_4), Ammonia (NH_3), Water (H_2O) and Hydrogen (H_2) and there was absolutely no oxygen.

They then passed electrical sparks through the mixture to signify lightning. The results were amino acids, the building blocks of proteins. It was later discovered that other energies like electricity, ultraviolet light, heat and shock also can excite gases and produce all 20 amino acids found in nature.

These twenty amino acids are found in all living organisms – plants animals and microorganisms and are the basis for their life, growth and reproduction.

This primordial soup was a diverse chemical environment containing a range of simple proteins (peptides) and fats (lipids). These constituents could have interacted with each other to produce more complex substances which formed the basis for the formation of life.

Ribonucleic Acids or some sort of its precursor could have been the first complex molecule to form in the course of 'Abiogenesis,' that is formation of life from inanimate things. RNAs are a type of simple biological material made of single strands of protein and containing nucleotides. Their special characters we will see later. This is also called the 'RNA World Hypothesis.'

Not everybody agrees with this hypothesis.

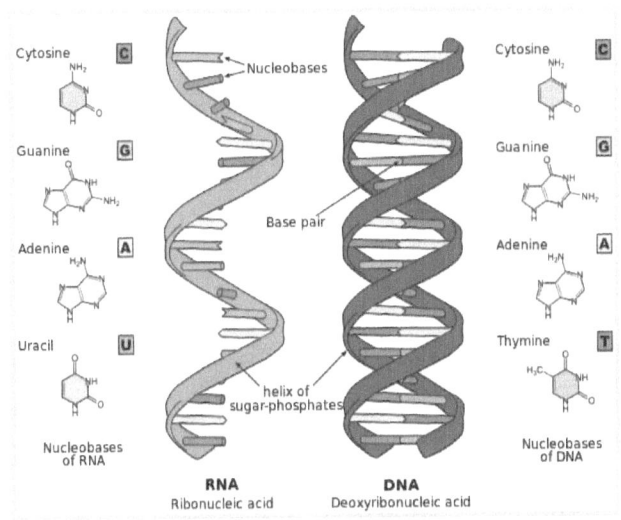

13. Picture showing the single stranded RNA and its related molecule, the DNA.

RNA was one of the earliest molecules which can replicate – which means it can create its own self again and again. This paved the way for organisms to reproduce themselves, the very essential character which differentiated the living from the non-living.

DNA is the molecule which is responsible for all life forms to transfer their genetic character to the next generation.

Some scientists propose 'Protein World Hypothesis' as proteins seem like a more natural starting point because they are easier to make than nucleic acids.

'GADV-Protein World' is a hypothetical stage of abiogenesis. GADV stands for the one letter codes of four amino acids, namely, glycine (G), alanine (A), aspartic acid (D) and valine (V), the main components of GADV proteins. In the 'GADV-protein world hypothesis,' it is argued that the prebiotic chemistry before the emergence of genes involved a stage where GADV-proteins were able to pseudo-replicate. It is also postulated that the information storage system found in the earliest rudiments of life would have been less advanced than the nucleic acid-based system

Reproduction

But how did the new life attain the most significant character of the living organisms, namely the miracle of reproducing their own like?

Ribonucleic Acids (RNA) have been part of the living organisms for billions of years, from the day they were created. RNA was also the primary living substance, largely due to RNA's ability to function as both genes and enzymes. They have the remarkable capacity of replicating themselves without help from other molecules. They take part in coding, decoding, regulation and expression of genes. These genes are the material responsible for transmitting the characteristics of the parent organisms to their off springs.

As for our narration, we could understand that whether it was RNA or some protein complexes to start with, they exhibited the miraculous power of replicating themselves from the molecular stages of formation of life itself.

This possibly sheds some light on the wonder called 'Reproduction,' which only living organisms exhibit.

From simple replication, reproduction has evolved over time into such an immensely complex process, even experts in the field are at a loss to understand the intricacies involved.

Early Life – The Unseen Micro Organisms

For the first three fourths of the earth's biological time (about 3 Billion years), microorganisms were the sole inhabitants of the earth.

As per the Virus-First Hypothesis, several investigators have proposed that viruses may have been the first replicating organic entities.

Most biologists now agree that the very first replicating molecules consisted of RNA, not DNA. It is possible today's single-stranded RNA viruses may be the descendants of these pre cellular RNA molecules.

Viruses are the smallest organisms which inhabit the earth. In fact they are so small, they cannot be seen even by ordinary light microscopes and one needs electron microscope to visualize them. They usually measure less than 200 nanometres,

Like other organisms, they possess genes, evolve by natural selection, and reproduce by creating multiple copies of themselves through self-assembly. But they do not have a cellular structure, which is often seen as the basic unit of life. Viruses are also described as 'Organisms at the Edge of Life' as some people describe them as organisms while some others call it organic structures that interact with living organisms.

Researchers Koonin and Martin (2005) state that viruses existed in this world as self-replicating units before the origin of cells. It is possible over time these units became more organized and more complex. Eventually, enzymes for the synthesis of membranes and cell walls might have evolved, resulting in the formation of cells.

The ancestors of modern bacteria were unicellular microorganisms which appeared on Earth about 4 billion years ago. They started their life, and survived through the Hadean eon, when the earth was in its formative stage. Hadean eon was named after the Greek god Hades, who was the king of the underworld and the god of death. It was named so because, during the eon, the earth was still hot, the rocks had not formed fully, there was tremendous volcanic activity and earth was undergoing heavy bombardment with celestial bodies. The microorganisms have proved themselves by living and thriving through the most adverse conditions on earth.

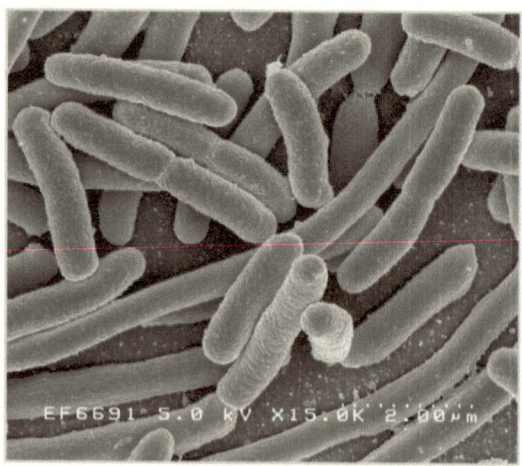

14. Bacteria – The oldest inhabitants of earth.

Electron micrograph of Escherichia coli bacteria grown in culture. E. Coli are present in good quantity in all human intestines.

The resilience of the bacteria can be seen even today. Scientists have discovered the bacteria *Pyrolobusfumari* inside a single hydrothermal vent in the Atlantic Ocean, 3,650 meters below the surface in temperatures up to 235 degrees Fahrenheit (113 Celsius). There are bacteria which can tolerate even higher temperatures. At present, there are also bacteria which can tolerate extreme cold, like the bacteria of the genus *Arthrobacter* and *Psychrobacter* .

Compared to the time the homosapiens have occupied this earth (0.2 million years), the bacteria had been here for eons longer, for about 4000 million years, or about 20,000 times the period of human existence. In short, they have proved their superiority for survival while we are yet to prove ourselves. As of now, we are in no way even remotely near to them in the survival strategies.

Plants, Our Parents

There was no free oxygen in this world when it formed about 4.5 billion years ago. The earliest organisms that appeared on earth were anaerobes, which means they did not depend on oxygen for their energy metabolism. Instead, they used sulphur, nitrates and fumarate for producing energy, which was technically inferior to the process that utilizes oxygen. The early organisms were alsoheterotrophs, which means they were are feeding on other organisms for nutrition.

Appearance of photosynthesis changed the entire scenario.

The evolution of photosynthesis started around 3.5 billion years ago. As the plants breathed in carbon dioxide which was freely available and breathed out oxygen, it led to a buildup of oxygen availability in this planet.

15. **Plants growing at the edge of water (Representative image).**

Several hundred million years ago, plants started growing at the edges of the water, and then out of it. Initially remaining close to the water's edge, mutations and variations resulted in further colonization of the land by the plants.

The first microbes to produce oxygen by photosynthesis were *oceanic cyanobacteria. They evolved into tufted microbial mats* more than 2.3 billion years ago. Initially, the free oxygen produced during this time was chemically captured by dissolved iron, converting iron {\displaystyle {\ce {Fe}}} in{\displaystyle {\ce {Fe^2+}}}to insoluble magnetite {\displaystyle {\ce {Fe^2+Fe2^3+O4}}} which sank to the bottom of the shallow seas to create massive, large scale, banded iron formations. The 'Great Oxygenation Event'uy6 started after these oxygen absorbers were filled to capacity and free oxygen started accumulating in the earth's atmosphere.

The increased production of oxygen set Earth's original atmosphere off balance and its biology into a catastrophe. Organisms inhabiting the earth until then had been living in an atmosphere of no oxygen. Most were obligate anaerobic organisms, meaning they could live only in an oxygen free environment. The rising concentration of oxygen was toxic to them and it destroyed most such organisms. This is called the 'Great Oxygenation Event' or the 'Oxygen Catastrophe.'

But even in this catastrophe, a few resistant forms survived and thrived. Some organisms developed the ability to use oxygen to increase their metabolism and obtain more energy from the same food. This also enhanced the development and nurture of complex multicellular organisms, including the Homo sapiens.

We also have important additional reasons to be thankful to our plants.

The oxygen produced on earth formed an ozone layer above the earth's atmosphere. It forms a powerful shield which protects the life on earth from the potentially very harmful ultraviolet radiation from the sun. The ozone layer absorbs 97 to 99 percent of the Sun's medium-frequency ultraviolet light (from about 200 nm to 315 nm wavelength), which otherwise would potentially damage exposed life forms on the earth's surface.

Now, we not only depend on the oxygen produced by the plants to breath, we also depend on them for our food. Humans have no way of producing energy giving food by themselves. They either have to eat plant food or eat animals, which again depend on plant food for survival.

It does look funny when people say they work to save plants and they are the guardians of the plant life.

Plants are out saviors and god fathers in more ways than one. Whether you destroy them relentlessly or try to nurture them, they nurture you and never abandon you. Of course, they never expect a 'Thank You' from you.

It is in your interest to nurture them and take care of them.

Last Universal Ancestor

It is believed that out of multiple protocells produced during the origin of life, only one line survived. Current evidence suggests that the last universal ancestor (LUA) lived about 3.5 billion years ago or earlier. This LUA cell is the ancestor of all life on Earth today. It was probably a prokaryote, which means lacking a nucleus or membrane-bound organelles such as mitochondria or chloroplasts, but possessing a cell membrane and probably ribosomes. Like modern cells, it used DNA as its genetic code, RNA for information transfer and protein synthesis, and enzymes to catalyze reactions. It is also possible that instead of a single organism being the last universal common ancestor, there were populations of organisms exchanging genes by lateral gene transfer, in fact making them our 'collective, ancestors.

The earliest cells absorbed energy and food from the surrounding environment. They used fermentation, which can happen in an oxygen-free (*anaerobic*) environment, to breakdown more complex compounds into less complex compounds. Fermentation produced only little energy, and the cells used the energy so liberated to grow and reproduce.

For major part of the world's history, that is for over three billion years, earth was inhabited by only microorganisms. Larger organisms came much later.

Several hundred million years ago, plants (probably resembling algae) and fungi started growing at the edges of the water, and then out of it. Initially remaining close to the water's edge, mutations and variations resulted in further colonization of the land. The timing of the first animals to leave the oceans is not precisely known: the oldest clear evidence is of insect like creatures with hard shell of a body and jointed legs (arthropods) crawling on land around 450 million years ago.

Around 380 to 375 million years ago, the first animals with four legs (tetrapods) evolved from fish. Fins evolved to become limbs which the first tetrapods used to lift their heads out of the water to breathe air. This would let them live in oxygen-poor water, or pursue small prey in shallow water. They may have later ventured on land for brief periods. Eventually, some of them became so well adapted to terrestrial life that they spent their adult lives on land, although they hatched in the water and returned to lay their eggs. This was the origin of the amphibians.

16. *Tiktaalikroseae*, a 375-million-year-old fossil animal, the link from water to land.

Tiktaalik represents an important step in the evolutionary transition from fish to animals that walked on land.

Discovered in Devonian-age rock on Ellesmere Island in Canada, the creature was a large aquatic predator with a flattened head and body.

The body plan and its environs suggest that it lived on the bottom in shallow water, and perhaps out of the water for short periods.

Terrestrial Animals

Land saw a vast array of animals traversing its surface from then on – tetrapods, arthropods, birds and what not. But surprisingly, an animal became bipedal, walking on two legs. The other two limbs were freed for work and with additional support from enhanced brain function, it started creating mischief all over the world. Paradoxically, it called itself 'Homo Sapien,' which in Latin means 'Wise man'!

The living organisms evolved and diversified so greatly over billions of years, the number of species living on earth right now is difficult to enumerate.

A fairly accurate evaluation conducted recently by the Hawaii's University, estimates that a total of 8.7 million species live on earth. But to date, only a total of 1.3 million species have been identified and described.

The predominant group is that of animals, representing 76% of all known species.

Within animals, arthropods are the group with the most species, with about 1.2 million species (1 million of which are insect species), representing 86% of all known animals. Our group, the chordates, is light years away from this figure, since it is made up of some 61,000 species (4% of the species),

Plants represent 17% of the species studied, with approximately 292,000 species.

More than 99 percent of all species, amounting to over five billion species that ever lived on Earth are estimated to have died out.

The living things of today present themselves in unimaginable varieties. They can be the giant Red Wood trees as tall as 300 feet, found in Sequoia National Park of California, U.S.A. or they can be so small that we cannot see them with our naked eyes and need a microscope to visualize them.

Their life span may be less than a few days as in Mayfly or Gastrorich or they may live for more than a few thousand years, as the Glass sponges

(Oopsacasminuta), the small cup like animals found in the South China Sea do.

But behind this apparent enormous diversity we see in this world, a common thread runs through every living thing. This unity of life we cannot discern, unless we go deep into their nature.

VIII. Unity of Lives

Do you have some carbon?

All living and non-living entities in this world are a combination of very basic substances called the elements.

Elements are chemically the simplest substances in this world and hence cannot be broken down using chemical reactions. Structurally, all of a particular element's atoms have the same atomic number.

As of now, the periodic table which scientifically tabulates the elements in this planet has 118 elements, Anything and everything in this world is made of one or a combination of these elements.

The ten most abundant elements in the universe in decreasing order of abundance are:

1. hydrogen

2. helium

3. oxygen

4. carbon
5. neon
6. nitrogen
7. magnesium
8. silicon
9. iron
10. sulfur

Humans

First we will take up the constitution of human body, an entity of our prime concern.

Almost 99% of the mass of the human body is made up of six elements: hydrogen (H), carbon (C), nitrogen (N), oxygen (O), calcium (Ca), and phosphorus (P). The next 0.75% is made up of the next five elements: potassium (K), sulfur (S), chlorine (Cl), sodium (Na), and magnesium (Mg). Only 17 elements are known for certain to be necessary to human life, with one additional element (fluorine) thought to be helpful for tooth enamel strength. A few more trace elements may play some role in the health of mammals.

The most abundant element by mass is oxygen. It accounts for nearly two-thirds of the mass of the human body. Carbon, hydrogen and nitrogen follow oxygen in the same order. In terms of number of atoms, hydrogen is the most abundant.

By mass, human cells consist of 65–90% water (H_2O), and a significant portion of the remainder is composed of carbon-containing organic molecules.

COMPOSITION OF HUMAN BODY

Element	Symbol	Percentage in Body
Oxygen	O	65.0
Carbon	C	18.5
Hydrogen	H	9.5
Nitrogen	N	3.2
Calcium	Ca	1.5
Phosphorus	P	1.0
Potassium	K	0.4
Sulfur	S	0.3
Sodium	Na	0.2
Chlorine	Cl	0.2
Magnesium	Mg	0.1
Trace elements include boron (B), chromium (Cr), cobalt (Co), copper (Cu), fluorine (F), iodine (I), iron (Fe), manganese (Mn), molybdenum (Mo), selenium (Se), silicon (Si), tin (Sn), vanadium (V), and zinc (Zn).		less than 1.0

Other Animals

Animal cells also broadly have the same proportion of the same chemicals as in humans.

Trees, Plants

Now the question arises whether the trees, which appear radically different living organisms have any similarity with us. We know they are also made of the elements found in this world. But we can see what they contain and in what quantity.

Living trees are very wet. Although there can be great variation between tree species, a living tree may be made up of more than two thirds water by mass. Thus, a living tree is made up of 15–18% carbon, 9–10% hydrogen, and 65–75% oxygen by mass.

Now, if we compare the molecular mass of human being and living tree, it goes like this:

	Humans	Trees
Carbon	18.5%	15–18%
Oxygen	65%	65 – 75%
Hydrogen	9.5%	9–10%
Water in cells	65–90%	80–90%

It is surprising how similar we are to animals and trees in our composition.

Amino Acids

Now we next come to the biological composition of the living organisms.

Amino acids are the building blocks of proteins, which in turn build our body. Throughout known life, there are 22 genetically encoded amino acids. These are the simple chemical building blocks on which the whole animal kingdom and the plant kingdom, including the 'proud and exemplary' homosapiens are built.

So, we humans share the same amino acids with the bacteria, trees, flowers, snakes, bulls and what not?

DNA

Let us proceed to more complex structures like DNA, which stands for Deoxyribo nucleic acid. These carry a copy of the organism's genetic material that is a complete blueprint of the organism. This DNA is present in the center of every animal cell, from jellyfish to you and me and in the center of every plant cell, from algae to orchids and in microorganisms. They transfer the genetic characteristics from one generation to the next. It determines that a child born to a human is human, a fly gives birth to a fly and an elephant's offspring is an elephant calf. It also ensures that a rose plant begets a rose plant and thorn bushes produce more thorn bushes.

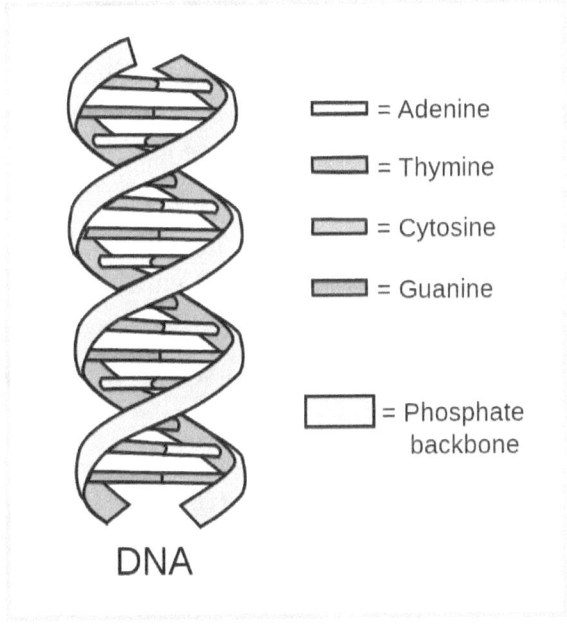

17. A DNA molecule. It is in the form of the famous "double helix" that looks like a twisted ladder.

There is only one type of DNA!

ALL animals and plants share the same DNA, which is basically composed of just four organic molecules. These are the vehicles which transmit the characteristic of any organism to its offspring.

At the chemical level, the cells of all plants and all animals contain DNA in the same shape – the famous "double helix" that looks like a twisted ladder. What is moresurprising is that, all DNA molecules – in both plants and animals – are made from the same four chemical building blocks, called nucleotides. They are adenine (A), cytosine (C), guanine (G), and thymine (T).

Then the next logical question arises. If we are so similar, how come we are different from an animal or a tree? It has been found that it is the sequence of these four afore mentioned nucleotides, and there by the information they encode in them, which decides what the organism is going to be. This sequence decides whether the organism will produce fins or wings, legs or stalk, tomatoes or potatoes.

So, there is only one type of DNA! ALL animals and plants share the same DNA which is basically composed just four organic molecules.

How much DNA do humans share with others

Humans are 99.9 percent similar to each other. 0.1% of the remaining genes tell us everything from our eye color, skin color, build etc. to whether we are predisposed to certain diseases like diabetes, hypertension or asthma.

Chimpanzees—our closest living evolutionary relatives—are 96 percent genetically similar to humans.

90 percent of the genes in the Abyssinian domestic cat are similar to humans.

Even bananas surprisingly share about 60 per cent of the same DNA as humans.

18. Just some changes in my DNA sequence...............and I could have been human!

An egret landing in water at Lake Pulicat, India.

Energy of Life

Almost all forms of energy available on earth are derived from the sun.

We are familiar with solar energy which can be tapped directly from the sun. The energy from coal, petroleum and natural gas has become indispensable to today's industries and our day today life. These are biomasses manufactured with sun's energy by organisms capable of photosynthesis and buried in earth millions of years ago.

Water on earth is evaporated with sun's energy to become clouds. When the ensuing rain water is stored in high mountains and comes down as rivers, that energy is used to produce hydroelectric power. Similarly, wind energy and its allied wave energy are produced by differential heating of the atmosphere by sun.

You can ask what about nuclear energy?

The radioactive heavy metals which produce nuclear energy were manufactured in the core of massive stars and scattered into the cosmic space when they exploded. Some of these metals were trapped by earth during its formation and that is what we are utilizing now for nuclear energy.

Growth, development, reproduction and every metabolic activity of all organisms, including the billions of people on earth, need a tremendous amount of energy. Again the ultimate source of all our energy is sun. Let us see how it works.

The source of the energy for all organisms is from what we call food or nutrients. Depending on how they get food, organisms are divided into autotrophs and heterotrophs.

Autotrophs are organisms that can produce their own food from the substances available in their surroundings using light (photosynthesis) or chemical energy (chemosynthesis).

Heterotrophs cannot synthesize their own food and rely on other organisms—both plants and animals—for nutrition.

Autotrophs produce their own food and energy by one of the following two methods:

Photosynthesis – Photoautotrophs use energy from sun to convert water from the soil and carbon dioxide from the air into glucose. Glucose provides energy to plants for their metabolic needs sustaining life. Glucose is also used by plants to make cellulose which is used to build cell walls and there by the structure of the plants.

Chemosynthesis – Chemoautotrophs use energy from chemical reactions to make food. The chemical reactions are usually between hydrogen sulfide/methane with oxygen. Carbon dioxide is the main source of carbon for Chemoautotrophs. Such bacteria are found inside active volcanoes, in hydrothermal vents in sea floor or in hot water springs. Such organisms are very few and exceptional and we can practically ignore them for the time being.

Photosynthesis is the primary process by which plants process light energy into chemical energy . This chemical energy is stored in carbohydrate molecules, such as sugars, which are synthesized from carbon dioxide and water. While carbon dioxide and water get recycled, energy from the sun gets trapped continuously by the process.

Plants are autotrophs and are called as a primary producer for the reason that they are capable of preparing their own food to gain energy.

Heterotrophs survive by feeding on organic matter produced by or available in other organisms. They get their energy by oxidation of preformed organic compounds, i.e. by eating other organisms either dead or alive.

So, sun is the ultimate source of energy for all life on earth and plants are the one and only intermediaries which can harness the energy and give it over to us.

Nearly all organisms use a process called Glycolysis to breakdown glucose molecule into two pyruvate molecules and store the energy released during this process as ATP and NADH. This pathway is common to both anaerobic and aerobic respiration.

All oxygen breathing organisms use the citric acid cycle, also known as the *TCA cycle* or the *Krebs cycle*, a series of chemical reactions, to release stored energy through the oxidation of acetyl-CoA derived from carbohydrates, fats, and proteins, into adenosine triphosphate (ATP) and carbon dioxide.

Adenosine triphosphate (ATP) is a complex organic chemical found in all forms of life. It provides energy to drive many processes in living cells. ATP is coined as the "molecular unit of currency" for intracellular energy transfer. When expended in metabolic processes, it converts either to adenosine diphosphate (ADP) or to adenosine monophosphate (AMP).

From the bacteria to the big animals or trees or human beings, everybody uses the same above chemical cycles to get energy.

Food Chain

The fact of dependence of all life on plants is clearly shown by what is called 'Food Chain.'

A food chain is the sequence of who eats whom in a biological community (an ecosystem) to obtain nutrition. A food chain starts with the primary energy source, usually the sun. The next link in the chain is an organism that makes its own food from the primary energy source—an example is plants that make their own food from sunlight using photosynthesis. These are called autotrophs or primary producers.

Next come organisms that eat the autotrophs; these organisms are called herbivores or primary consumers—an example is a rabbit or goat that eats grass.

Secondary consumers eat primary consumers. They are carnivores (meat-eaters) and omnivores (animals that eat both animals and plants).

In turn, these animals are eaten by larger predators called tertiary consumers.

These animals can be eaten by quaternary consumers

Each food chain ends with a top predator and animal with no natural enemies. They are called apex predators – examples being a lion, eagle or a king cobra.

A network of many food chains is called a food web.

The top most apex predator is the human being, who keeps eating plants and animals at his will and pleasure. He is also the biggest threat to our ecology, having destroyed almost half the animal and half the plant population the world should have had by now in its natural course.

How Come we Dominate Other Species?

When we are chemically, biologically and structurally so similar to other lives on earth, how is that humans lord over all other species on earth and established themselves as the prime species of the planet?

19. A group of humans hunting the massive Mammoth.

Humans did not evolve massive bodies or big fangs or huge claws for their survival. Instead, they evolved the ability to bond into communities and to cooperate with one another. This strategy turned out to be far more powerful than any other animal which inhabited earth.

Humans did not evolve massive bodies or big fangs or huge claws for their survival. Instead, they evolved the ability to emotionally bond into communities and families where they became largely inclined to cooperate with one another. These communities and families which worked together turned out to be far more effective and powerful than any other beast which inhabited earth.

It helped us survive and thrive. This is not exactly romantic or sexy, but it is true.

This has brought us to a point where we have to examine the bonding we humans experience. Let us start from the most basic drive for bonding in the animal world, the sexual, and then go on to the more sophisticated forms of bonding – social and spiritual.

IX. Origin of Sex

Birth of the Original Sin

Question of Sex

Sex is a universal phenomenon, occupying every nook and corner of the world – the very reason why you, me and everyone is born. Such an important entity as this, it definitely deserves a deeper and an elaborate analysis.

The first question which raises its head is – is there any need for this enigmatic enterprise called sex, detested as 'sinful' by major religious disciplines, relegated to darkness and secrecy by the society and even a discussion on the topic deemed indecent by the 'decent'?

We will start from the possible origin of 'sexual reproduction,' which is coming together of two organisms to produce more such organisms. We will also see whether it has any purpose and what role it has in the grand scheme of things.

'Union' Even Before Origin of Life

Scientists assume life originated from organic molecules which were formed in a 'primordial soup.' The molecules swimming in early Earth's primordial soup would have been continually destroyed by ultraviolet radiation from the sun, as well as heat and other processes on the planet.

But when certain special pairs of molecules combined to form a larger compound, they sometimes came out with protections that neither had alone.

One example is the compound of glutamic acid and two glycine molecules.

Individually, each of these molecules was easily destroyed by ultraviolet radiation. But put together, they were extremely stable.

So, combining of molecules to withstand the onslaught of nature and surviving may be the earliest example of the survival by unity.

Sex Originated with Life

The first living cells wereunicellular organisms that did not have a membranebound nucleus, mitochondria, or any other membrane bound micro structures inside their cells and are named prokaryotes. Species with nuclei and organelles inside their cell are placed in the domain, Eukaryota, under which all multicellular organisms, including humans come.

Prokaryotes reproduce without fusion of gametes and are divided into two domains, Archaea and

Bacteria. Archaeal cells have unique properties separating them from the other two domains of life, Bacteria and Eukaryota.

Despite their morphological similarity to bacteria, archaea possess genes and several metabolic pathways that are more closely related to those of eukaryotes, notably for the enzymes involved in transfer of genetic material. In a way, we can say archaea are midway between the prokaryotes and the eukaryotes.

Scientists say archaeal ancestor may already have had DNA repair mechanisms based on DNA pairing and recombination and possibly some kind of cell fusion mechanism. There was acrisis for survival and a desperateneed for such reparation during the 'Oxygen Catastrophy,' a time when the atmosphere of the earth changed from 'oxygenless' to the 'oxygen rich' due to the oxygen produced by the plants. Oxygen is a highly reactive element and it proved highly damaging to the vast majority of organisms which were not dependent on oxygen for their survival and had no exposure to oxygen at all. There was almost a total annihilation of all living organisms due to oxygen exposure and very few species survived and adopted.

The detrimental effects of oxygen and their derivatives on the archaealgenome could have promoted the evolution of meiotic sex, which is the basis for sexual reproduction. In meiosis, the genetic material and chromosomes of one cell divide into half. It then fuses with another cell of opposite sex which has similarly become half. Thisunion completes the genetic configuration of the new offspring which they produce together. Selective pressure for efficient DNA repair caused by oxidative DNA damages may have driven the evolution of eukaryotic sex on similar lines, involving such features as cell-cell fusions, cytoskeleton-mediated chromosome movements and emergence of the nuclear membrane.

"Libertine Bubble Theory"

If you think being loose and unrestrained in sexual activities is a phenomenon of the modern times, you are mistaken. A scientist says such activity could have occurred with the origin of life, in the very early organic molecular stage.

Thierry Lode, a French biologist and a Professor of Evolutionary Ecology at the University of Rennes in France has proposed a theory where the evolution of sex can alternatively be described as a kind of gene exchange that is independent from reproduction. According to Thierry Lodé's "Libertine Bubble Theory," sex originated from an archaic gene transfer process among prebiotic bubbles. The contact among the pre-biotic bubbles could, through simple food or parasitic reactions, promote the transfer of genetic material from one bubble to another. That interactions between two organisms be in balance appear to be a sufficient condition to make these interactions evolutionarily efficient, that is to select bubbles that tolerate these interactions ("libertine" bubbles) through a blind evolutionary process of self-reinforcing gene correlations and compatibility.

The Libertine Bubble Theory also states that current sexual species could be descendants of primitive organisms that practiced more stable exchanges in the long term, while asexual species have emerged much more recently in evolutionary history.

Thus the evolution of sex and evolution of the organisms were likely inseparable processes that evolved in large part to facilitate DNA repair. Constant pressure to repair itself against damages has been proposed to explain the universal maintenance of meiotic sex in eukaryotes. In eukaryotes, it is thought to have arisen in the Last Common Eukaryotic Ancestor (LECA) about 1.2 billion years ago, possibly via several processes of varying success, and then to have persisted.

So sex, or some form of interaction between two organic entities whence they blend and also exchange their genetic material, seems to have been present from the time life originated. Possibly it started as a repair and maintenance service. But over time, it became the one which has sustained the life over our planet, enriched and diversified life in innumerable ways and made it complex beyond anybody's imagination. From the time it originated in 'unconscious,' non-thinking, un-aware organisms to the modern 'thinking,' 'conscious' human beings, it has manifested in innumerable ways.

Coming together of the sexes, so that two gametes meet and form a zygote which produces an offspring, is a simple straight forward process which happens naturally throughout the animal kingdom.

But the 'thinking man' has made the simple procedure extremely complicated, with heavy emotional and psychological investments in them. It has become so complex, people have no idea how to manage them. The experts who call themselves psychologists and sociologists who are supposed to understand these things better, seem to create more confusion than clarity in the common man.

The main problem seems to be the concept of 'Love,' which entity has made all human bonding complex and a source of eternal conflicts.

Without digressing further, we will return to the multicellular organism of our primary concern, namely human beings to find out how they handle it.

X. Origin of Love

How the babies taught us to be intelligent and responsible!

The evolutionary processes have sculpted not merely our body, but also the brain, the psychological mechanisms it houses, and the behavior it produces. Love too has evolutionary history of millions of years.

Psychological adaptions, according to evolutionary theory, are mechanisms our species develops to solve problems. These adaptions are passed on to succeeding generations if they contribute to our survival and reproduction, the two ultimate concerns of any living organism.

The lower animals do have affinity to each other and certain amount of co-operation among themselves. Charles Darwin, in his book 'The Descent of Man' (1871) says – 'Firstly, the social instincts lead an animal to take pleasure in the society of its fellows, to feel a certain amount of sympathy with them, and to perform various services for them. The services may be of a definite and evidently instinctive nature; or there may be only a wish and readiness, as with most of the higher social animals, to aid their fellows in certain general ways.'

Darwin also says, '............that any animal whatever, endowed with well-marked social instincts, the parental and filial affections being here included, would inevitably acquire a moral sense or conscience, as soon as its intellectual powers had become as well, or nearly as well developed, as in man.'

From an evolutionary perspective, Darwin implies morality as a form of cooperation and a concern for the well-being of the fellow organisms. Cooperation requires individuals either to suppress their own self-interest and / or cater to the interest of others also.

Archaeological evidence supports the emergence of such moral reputation as a key driver of selection from early human evolutionary history. Selection for pro-social intentions and behaviors occurs from at least 1.5 million years ago, with increasing material evidence for care of the vulnerable as well as displays of positive reputation in material culture. Complex social dynamics similar to those of modern hunter-gatherers and associated with collaborative morality become identifiable by at least 100,000 years ago, which corresponds to a period associated with the expansion of the brain of the homosapiens.

The final acceptance of our evolution *from apes* might have been a significant point. But there is as yet no real answer as to how we evolved. What had happened *in between* ancestral apes and humans? Why are all the other apes still eating leaves and picking parasites off each other while we sit in tea shops making conversation on politics or arts, or travel around the whole world collecting more and more knowledge? What spark made us human, and how did it occur?

More than learning how to use tools, more than being successful at violence, more than adapting to moving out of the forest into the grasslands of Africa, it was learning how to love and live with each other that contributed to human survival and domination of the world. It also drove the human evolution!

But how did this enigma called love happen?

How did we evolve the most loving and most powerful brain on the planet?

As usual scientists have several theories. One of the major contenders is related to the helplessness of the human new born and the efforts needed to nurture it.

'Useless Baby'

No other animal on earth delivers such helpless, 'useless' and demanding babies as the humans.

There is an interesting reason for the same.

As our brains grew, the size of the heads of the babies also grew. But there is a limitation to the size of the hip (pelvic) bone which houses the birth canal of the mother. The baby has to come through this narrow birth canal, enclosed by pelvic bones, to see this world. With this constraint, our babies had to be born earlier in development, before full maturity with softer and smaller heads. With full growth, their heads would be too big and too hard to pass through the birth canal and may cause grave injuries with devastating consequences to the mother and the baby.

Larger brains led to more premature birth. This called for nurturing of the new born for a much longer time. Longer duration of care and nurture demanded more skills and intelligence on the part of the care givers which made the brain grow even bigger. As fully grown bigger heads cannot pass through the woman's birth canal, babies were delivered in a more premature form, before the full growth of the brain. In this vicious cycle, the brain size of the humans had a great boost to grow bigger and bigger and as an offshoot, birth of more and more premature babies who needed much longer duration of care.

20. **Rearing the vulnerable children needed a family and the family needed the society for its survival.**

The growing length of childhood co-evolved with the enlarging of the brain. Our brain has tripled in size over the last 2.5 million years, since the time of the first tool-making hominids. Another contributing factor to the brain growth is the development of complex bonding the animals developed among themselves, which was necessary for the protection and care of the helpless new born. Thisbonding included friendship, love, parent – child attachment and loyalty to a group.

In humans, there are additional reasons for the childhood to be long. The human child has so much to learn by way of skills and knowledge which have been accumulated over generations, a burden which no other animal has. To keep a vulnerable child alive and to groom it for many years, we evolved strong bonds between parents and children, between mates, within extended family groups, and within bands as a whole – all in order to sustain the new born which needed all the care available in this world to survive and thrive. Bands with better teamwork excelled over other bands for scarce resources. Since breeding occurred primarily within bands, genes for bonding, cooperation, and altruism proliferated within the human genome.

This prolonged childhood created a new risk.

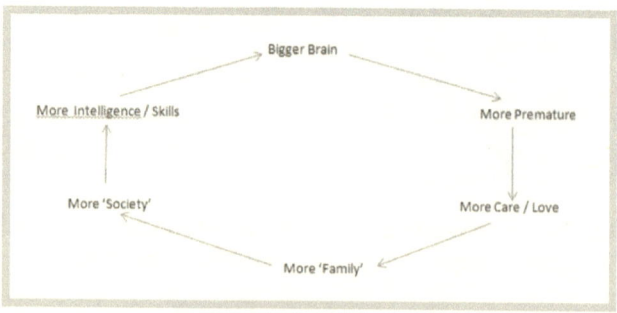

21. Vicious Cycle of Love promoting brain and Brain promoting love.

In many primates today, a mother with a dependent infant is unavailable to mate until her infant is weaned. To get access to her, a male would first have to kill her child. This sort of targeted infanticide goes on in many species, including gorillas, monkeys and dolphins.

This led Kit Opie of University College London in the UK and his colleagues, to propose a startling idea. Almost a third of primates form monogamous male-female relationships, and he suggested that this behavior had evolved to prevent infanticide.

In those species where males and females started bonding strongly, their offspring's chances of survival improved because the males could help out with parenting. As a result, monogamy was favored by evolution, says Opie.

Extra male care helped early human societies grow and thrive, which in turn allowed our brains to grow larger than our closest animal relatives.

There is evidence to back this up. As brain size started to expand, so did cooperation and group size. We can see a trend towards larger groups and more cooperation in the early-human species as their brain sizes increased.

Another supportive finding is that regions of the brain that deal with love have only appeared quite recently in our evolutionary history.

There is an alternative theory, though not so popular as the one mentioned before, which suggests that monogamy evolved as an outgrowth of a "mate guarding strategy": that is, males staying with a female to ensure that no one else mates with her. We need not go into controversial details of the theory now. May be jealous husbands of today can shed more light on how effective the strategy is.

But there is one undeniable trait all human ancestors have in common – a strong mother-child bond.

This brings us to Sussman's suggestion that romantic love evolved out of mother-child bonding. The bonds of long-term couples are similar to those between mother and child, and rely on similar hormonal processes.

Whether it was infanticide or a mother's attachment to her infant that pushed us to bond closer, we should be thankful that something did. We owe much of our success as a species to that crazy thing called love.

Archaeology of Love

Dr. Spikins, the author of a book named 'How Compassion Made Us Human' explains why altruism was central to human evolutionary origins. She says "The traditional view sees Neanderthal childhood as unusually harsh, difficult and dangerous. This accords with preconceptions about Neanderthal inferiority and an inability to protect children epitomizing Neanderthal decline. Our research found that a close attachment and particular attention to children is a more plausible interpretation of the archaeological evidence, explaining an unusual focus on infants and children in burial, and setting Neanderthal symbolism within a context which is likely to have included children."

In 1908, Bouyssonie brothers, both of them being archeologists, uncovered the 50,000-year-old Neanderthal skeleton in the cave at La Chapelle-aux-Saints, France. Almost immediately they speculated that the remains were intentionally buried.

The skeleton discovered by them belonged to a Neanderthal who was missing most of his teeth and showed signs of hip and back problems that would have made movement difficult without assistance. Evidence from the La Chapelle site suggests that Neanderthals were like us in that they cared for their sick and elderly.

"Before they took care of his dead body, the members of his group would have had to take care of him while he was alive," archaeologists surmise.

The 'First Family'

In 1976, Johanson and White discovered the fossilized remains of Australopithecus afarensis, the Hominidae, who lived 3.75 million years ago, in Hadar, Ethiopia. These fossils were found to be a lineal ancestor to the human race, and were named the "First Family," because the fossils of 13 individuals (both sexes, including children) were excavated from the same ruins. However, the first family's emotions, behaviors and societies are not fossilized and unfortunately, we have no way of knowing them.

Love seems to have a prehuman origin and seems to have been handed over to us over millions of years. It presents itself as a fact of nature.

We, as modern people, always have the tendency to meddle with nature, complicate things and land into chaos. No wonder, love also gets tampered with!

As of now, love in this planet seems to present itself in varied forms. But the most common form of love, which occupies a large part of a person's mind and most perceptible in the modern times is the so called 'Romantic Love.' So, let us start with it to explore what love means to the human beings.

XI. Romantic Love

*I know I am in love with you because my reality
is finally better than my dreams*

– Dr. Seuss

Obviously, the most visible bonding on earth is the so called 'Romantic Love.' It is available everywhere – on television, movies, internet, magazines, newspapers, advertisements and also live on the streets. People believe the most common prelude to the coming together of sexes is 'Romance.'

Most people have been conditioned to believe that so-called romantic love is the most important pursuit in life and that only the ones who have found it are truly happy and fulfilled.

22. The Ethereal world of Romance.

A Japanese pair going through the pre-matrimonial photo shoots.

The belief that we are not complete as we are and that we need someone else to fill in our sense of existential emptiness is not an urge of the day. In fact, it goes back to ancient times.

Plato's 2,500 year old philosophical text 'Symposium' describes the origins of humanity, through the words of Aristophanes, a playwright of his times. It says the original form of man was a four-legged, four-armed, double-sexed entity.

Not only did early humans have both sets of sexual organs, but they were outfitted with two faces, four hands, and four legs. These monstrosities were very fast – moving by way of cartwheels – and they were also quite powerful. So powerful, in fact, that the gods were nervous for their dominion.

Wanting to weaken the humans, Zeus, Greek king of Gods, decided to cut each in two, and commanded his son Apollo "to turn its face…towards the wound so that each person would see that he had been cut and keep better order." If, however, the humans continued to pose a threat, Zeus promised to cut them again – "and they'll have to make their way on one leg, hopping!"

The severed humans were very miserable and *"Each one longed for its other half, and so they would throw their arms about each other, weaving themselves together, wanting to grow together."*

Finally, Zeus, moved by pity, decided to turn their sexual organs to the front, so they might achieve some satisfaction in embracing.

Plato again quotes Aristophanes to explain "the source of our desire to love each other." He says,

"Love is born into every human being; it calls back the halves of our original nature together; it tries to make one out of two and heal the wound of human nature. Each of us, then, is a 'matching half' of a human whole…and each of us is always seeking the half that matches him."

Hinduism also has a beautiful myth about the origins of love. In the beginning, there was a super being called Purusha. This being was without desire, craving, fear, or indeed the impulse to do anything at all—since the universe was already perfect and complete.

Then the creator Brahma took out his divine sword and split Purusha in two. Sky became separate from earth, darkness from light, life from death and male from female. Each of these set off passionately to reunite with its severed half. Though we could not surmise much about heaven and earth, the craving between men and women is there before us, day in and day out.

We are in an endless search for that "special someone"—the soul mate—the universe created just for us, and we are willing to give up anything to "lose ourselves" in his or her embrace.

The moment we fall in love, the world turns into a magical place. Suddenly, life becomes more beautiful, adventurous, meaningful…in short, life becomes worth-living again. But having experienced the emotional high of romantic love, we want more of it. People start behaving like addicts: have obsessive thoughts, participate in risk-taking activities and find it hard to deal with withdrawals. It is so overwhelming that it distorts our perception and often leads us into making choices that we will later regret.

There is a biological explanation for all these phenomena. The brain of people in love secretes an ample amount of hormones like dopamine, norepinephrine and serotonin—hormones that boost the pleasure and confidence levels. Under this hormonal boost, everything seems perfect, blissful and heavenly to the romantic love addict. Yet, once the initial high fades, just as the wearing out of the 'high' of alcohol or psychedelic drugs which ultimately lead to the depression of senses, everything changes. Life becomes mundane, ordinary, boring once again. In the long run, the illusions disappear and rigors of everyday life open their eyes to the harshness of reality. Once that happens, the inner emptiness returns and torments him, as much or even more than earlier.

An Assyrian proverb, coined several thousand years ago says it all – 'Love and eggs are best when they are fresh.' Somehow, it is consoling to know people have undergone such heart breaks over thousands of years.

The story of Romeo and Juliet does not seem to be a celebration of love. They appear more like warnings as to how romantic love can potentially ruin everything. So are the travails of Laila – Majnu, Shirin – Farhad or the south Indian legend of Ambikapathy – Amaravathy. The love story of Paris and Helen of Troy is told in the Homerian epic 'Iliad.' The word Iliad translates into ' a long series of woes and disastrous events.' Death, destruction and humiliation seem to be the outcome of all these legends. Remarkably, none of them end up in successful union. Possibly, that is why they still remain as dream legends, as a union might have led to domestic quarrels and drown them into the dirty burdens of everyday life.

To escape the trap of romantic love is quite difficult, considering that we have billions of dollars being spent by commercial interests and entertainment industry which promotea misrepresentation of actual love relationships. In our consumer oriented commercial world, anything and everything can be harnessed to make money. So our business establishments exploit our deep rooted emotional need to love and be loved. They use romantic love as a bandwagon to push their products to the people at large. We see love overflowing, dripping and flooding when a woman is presented with an exotic diamond necklace. We see it germinating in high end branded clothes and flowering in five star hotels.

23. An advertisement for jewelry.

Romantic love needs additional reinforcements of several kinds and some of them come at a hefty price.

Love on Fast Track

In these days of fast cars and fast food, everybody is in a hurry and people want even the love to be fast.

There are ample business minds around which quickly perceived the opening it offered to make a fast buck and have started organizing events called 'Speed Dating.' Now it has become a popular event in many parts of the world, where males or females who were not able to bond with suitable mates in their normal course of life meet a good number of partners in a short time. Within a period of about three to ten minutes of interaction, they have to decide which partner they like and whom they dislike. Speed Dating just seems to be a manifestation of frustration our modern men and women suffer in acquiring a 'Dream Partner.'

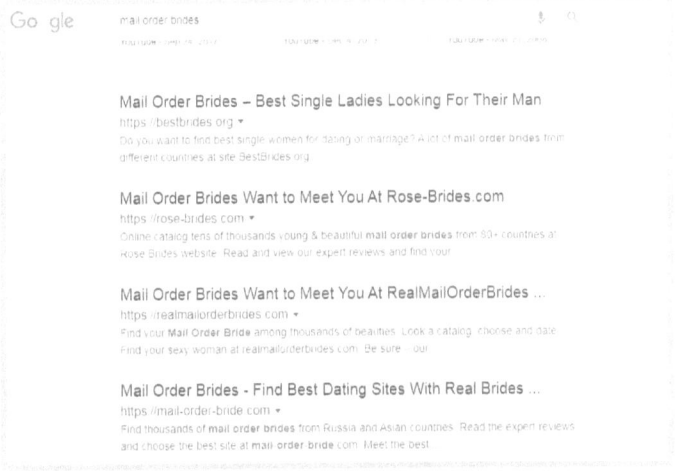

24. Mail Order Brides.

A screen shot of an on line search for Mail Order Brides. Your partner may be just a click away. But there is no assurance for the durability or compatibility of the package.

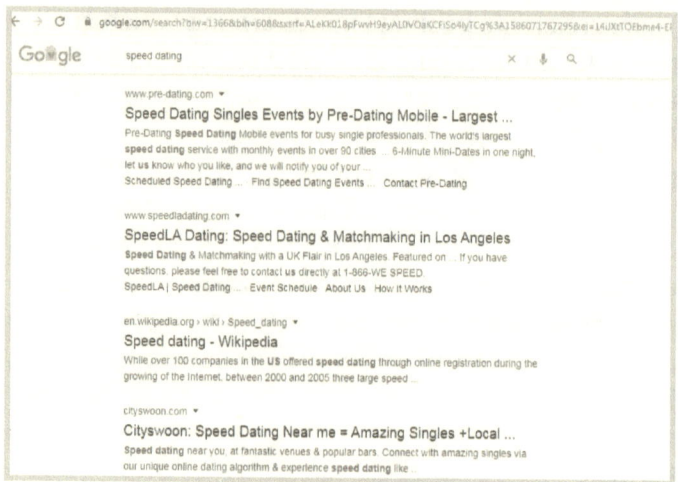

25. Speed dating

A screen shot of on line search for speed dating sites.

If you have no time even to dicuss your likes and dislikes with your sweet heart, no worry. You can do everything in a whiff.

After all, life is short and we can not spend too much time haggling on small personal details!

Another manifestation of this social problem is the so called 'Mail Order Brides.' In this, usually men from developed countries get access to women in 'not so developed' countries through commercial agencies. The women are usually the ones who are eager to immigrate to better countries. These agencies arrange for everything from making the likely couple meet each other to immigration of the 'bride' to the developed country.

Commerce and love are very strange bed mates and no wonder love never thrives in its company. But business thrives and so, it is advantage business.

Romantic love of today is most often immature and superficial. If we dissect deep into them, we usually find there is interplay of infatuation, sexual drive, greed, possessiveness and ego, masquerading as love of the highest order.

People often get into toxic relationships where they do not love each other, but they love the idea of each other. They are in love with the fantasy that is constantly playing out in their head. And instead of ditching the fantasy and getting on with the person in front of them, they spend all of their will and energy into tormenting the person to fit the fantasy image they nurture. No wonder the home is not filled with happiness, but anger and frustration!

What is the solution?

We need to understand that there's no person out there who will complete us.

We will go on experiencing inner emptiness no matter how many romantic love partners we have.

Unless we are able to love and accept ourselves, we will always feel incomplete, regardless of how much love is shown to us.

Healthy love relationships can only exist between two already complete, fulfilled people who decide to enrich their lives by sharing their love. Sadly, most of us are emotional beggars seeking for completion in another person, only to find that our emptiness is actually deepening once we wake up to the fact that our partner is just as empty as we are.

We need to understand that love relationships are based on freedom.

Normally, partners are trying to fit each other into their fantasy image of how the Ideal Lover should be, depriving them of their freedom to be themselves. They act out their lives in the drama of the world, with no life or soul in it.

Just like birds need freedom of space so that they can soar into the skies, partners in an intimate relationship need the freedom to be themselves so they can spread their wings of love and fly to the peaks of consciousness.

There is no great relationship without a lot of work behind it. In other words, great relationships do not happen to be—they are slowly built with care and effort.

Contrary to what the romantic tradition has made us believe, the Perfect Relationship does not exist other than in our imagination.

Does it mean we are at a dead end as far as love goes?

No. Definitely not!

Love goes far beyond the trappings of romantic love and we will presently see the different types of love the world has seen, from ancient times to up till now.

XII. Different Types of Love

I was in love with loving.

– Saint Augustine, Confessions

Humans had the necessity to live together and prosper together. Love among them was an existential fact.

But all loves are not same and the manifestations they take can ne innumerable. Being aware of this from time immemorial, the ancient Greeks have described eight types of love. They are:

Eight Types of Love – Greek

1. "Eros" *or Erotic Love*

The first kind of love is *Eros*, which is named after the Greek god of love and fertility. Eros is a passionate and intense form of love that arouses romantic and sexual feelings.

The ancient Greeks considered Eros to be dangerous and frightening as it involves a "loss of control" through the primal impulse to procreate.

Eros is a primal and powerful fire that burns out quickly. It needs its flame to be fanned through one of the deeper forms of love below as it is centered around the selfish aspects of love, that is, personal infatuation and physical pleasure. We can also say it is mediated by the physical body

2. "Philia" *or Affectionate Love*

The second type of love is *philia,* or friendship. The ancient Greeks valued philia far above eros because it was considered a love between equals.

Plato felt that physical attraction was not a necessary part of love, hence the use of the word *platonic* to mean, "without physical attraction." Philia is a type of love that is felt among friends who've endured hard times together.

As Aristotle put it, philia is a *"dispassionate love"* that is free from the intensity of sexual attraction. It often involves the feelings of loyalty among friends, camaraderie among teammates, and the sense of sacrifice for your pack.

3. "Storge" *or Familial Love*

Although *storge* closely resembles *philia* in that it is a love without physical attraction, storge is primarily to do with kinship and familiarity. Storge is a natural form of affection that often flows between parents and their children, and children for their parents.

Storge love can even be found among childhood friends that is later shared as adults.

4. "Ludus" *or Playful Love*

Although *ludus* has a bit of the erotic *eros* in it, it is much more than that. The Greeks thought of ludus as a playful form of love, for example, the affection between young lovers. It may be in the early stages of falling in love with someone, which may involve flirting, teasing, and feelings of euphoria. Yet playfulness is one of the secrets to keeping the childlike innocence of your love alive, interesting and exciting.

5. "Mania" *or Obsessive Love*

Mania love is a type of love that leads a partner into a type of madness and obsessiveness. It occurs when there is an imbalance between *eros* and *ludus*.

In manic love, a person is desperate to love and be loved. Because of this, they can become possessive and jealous lovers. They even resort to abnormal and extreme measures in an attempt to achieve their end.

If the other partner fails to reciprocate with the same kind of *manic* love, it may lead to psychological and social issues and conflicts with law. The murders, suicides and other violent incidents we come across presently may be outcome of this form of love.

6. "Pragma" *or Enduring Love*

Pragma is a love that has aged, matured and developed over time. It is beyond the physical, it has transcended the casual, and it is a unique harmony that has formed over time.

You can find *pragma* in married couples who have been together for a long time, or in friendships that have endured for decades. Unfortunately pragma is a type of love that is not easily found. We spend so much time and energy trying to find love and so little time in learning how to maintain it.

Unlike the other types of love, pragma is the result of effort on both sides. It is the love between people who have learned to make compromises, have demonstrated patience and tolerance to make the relationship work.

7. "Philautia" *or Self Love*

The Greeks understood that in order to care for others, we must first learn to care for ourselves. This form of self-love is not the unhealthy vanity and self-obsession that is focused on personal fame, gain and fortune as is the case with Narcissism.

Philautia is self-love in its healthiest form. It shares the Buddhist philosophy of "self-compassion" which is the deep understanding that only if you have the strength to love yourself and feel comfortable in your own skin, will you be able to provide love to others. As Aristotle put it, *"All friendly feelings for others are an extension of a man's feelings for himself."*

You cannot share what you do not have. If you do not love yourself, you cannot love anyone else either. The only way to truly be happy is to find that unconditional love for yourself. Only when you learn to love and understand yourself, will you be ready to search for the spiritual freedom of the Self.

8. "Agape" *or Selfless Love*

The highest and most radical type of love according to the Greeks is *agape,* or selfless unconditional love.

This type of love is not the sentimental outpouring that often passes as love in our society. It has nothing to do with the condition-based type of love that our sex-obsessed culture tries to pass as love.

Agape is what some call spiritual love. It is an unconditional love, bigger than ourselves, a boundless compassion, an infinite empathy. It is what the Buddhists describe as *"mettā"* or "universal loving kindness." It is the purest form of love that is free from desires and expectations, and loves regardless of the flaws and shortcomings of others.

Agape is the love that accepts, forgives and believes for our greater good and borders on divinity.

Five Types of Love in Ancient India

Hinduism teaches that love consists of a series of stages through which an individual can climb. The lower stages on love's journey are not necessarily supposed to go away as one gets more enlightened. However, remaining stuck at the lowest rung can cause a lot of frustration and sadness.

Here is a look at the five stages of love as the Hindus see them:

1. "Kama," or Sensory Craving

At first, the desire to merge gets expressed through physical attraction, or kama. Technically speaking, kama means "craving for sense objects," but it is usually translated as sexual desire.

In ancient India, sex was not associated with shame—as shown in many interpretations of the Judeo-Christian myth of the "fall of man"—but a joyous aspect of human existence and a topic worthy of serious investigation. The Kama Sutra, which was written around the time of Christ, is not merely a catalogue of sexual positions and erotic techniques. The majority of the text is a philosophy of love dealing with questions such as what sparks desire, what maintains it, and how it can be wisely cultivated.

But while the sages honor kama as a legitimate goal of life, they insist that we will never achieve wholeness through the act of sex alone.

2. "Shringara," or Rapturous Intimacy

Sex without true intimacy and sharing can leave us feeling empty. That is why the philosophers of India focused on the emotional content of the experience.

Out of this bubbling stew of feelings is born shringara, or romance. Lovers stir the pot of their erotic attraction by seeing one another as embodiments of all their cravings. And they spice it up by sharing secrets, making up affectionate names for one another, playing games, and giving inventive gifts.

The ancients were realistic about what a mixed bag romance is. They did not imagine that finding our "soul mate" would solve our problems, relieve our sense of unworthiness and self-doubt, or satisfy all of our emotional needs.

Nevertheless, Indian philosophy teaches that romantic love, enjoyed in moderation, provides a foretaste of something even greater.

3. "Maitri," or Generous Compassion

Compassion resembles the uncomplicated love that we naturally feel toward children and pets. It is also associated with "matru-prema," the Sanskrit term for motherly love, which is said to be love's most giving and least selfish form. Maitri is like a mother's tender love expressed towards all living beings and not just for one's own biological child.

4. "Bhakti," or Impersonal Devotion

Beyond interpersonal love, the Indian tradition envisioned an impersonal form in which our sympathies gradually expand to embrace the whole of creation.

It can be translated as the cultivation of the self through the love of God. Luckily for those who are not conventionally religious, it can be directed toward whatever ideal we hold on a high pedestal, like love, kindness, truth, or social justice.

Through 'bhakti' one learns that one's true family is the family of life itself.

5. "Atma-Prema," or Unconditional Self-love

Up until this point, each stage of love has been directed outward into the world. But at its apex it comes full circle back to the self. Atma-prema can be translated as "self-love." This is not the self as we usually think of it, but the essential self, the self that exists at the center of all of us.

What this means in practice is that we see ourselves in others and see others in ourselves. "The river that flows in you," says the Indian mystical poet Kabir, "also flows in me." When we are swept by that flood, we recognize that we are all expressions of the One Life.

Atma-prema arises from the realization that beyond our personal faults and imperfections—beyond even our name and personal history—we are all children of the most high. When we love ourselves and others in this profound yet impersonal way, our love becomes boundless and unconditional.

The sages of ancient India did not view the five stages on love's journey as mutually exclusive. You do not need to renounce sex and romance in pursuit of a "higher" love. All the forms can coexist in the heart that is mature.

We do not live on the summit of universal love. We have to make arduous efforts to climb the mountain, step by step, and then we descend it in order to share what we have found with others.

By fully developing each of the five stages in turn, we can free ourselves from the addiction to romance as well as our narrow attachment to kith and kin, and become transformed into genuinely full-hearted people.

XIII. Mature Love

*I, you, he, she, we—
in the garden of mystic lovers,
these are not true distinctions.*

– Rumi

The deepest need of man is the need to overcome his separateness, to leave the prison of his aloneness. Man, of all ages and cultures has confronted the same challenge – how to overcome separateness, how to achieve union, how to transcend one's own individual life and find some fulfillment.

People have tried various methods to transcend this loneliness – by wielding absolute power over others, sadistic pursuits to break into a person's integrity and umpteen other pursuits. But the only healthy and humane possibility for achieving this is through love.

To come to basic questions, what is love?

Love is the active concern for the life and the growth of that which we love.

There are some basic elements common to all forms of love. These are care, responsibility, respect and knowledge.

Without the above elements, mature love cannot happen.

The only way a person can achieve the capacity to have mature love is to develop his total personality. A person has to have true humility, courage, faith and discipline to be capable of mature love. One has to deliberately and steadfastly cultivate these characters to be capable of experiencing a mature, fulfilling love. In a culture in which these qualities are rare and not valued, the attainment of the capacity to love also seems to be a rare achievement.

It takes a 'complete' person to enjoy wholesome love. When we are incomplete, we are always searching for somebody to complete us. When, after a few years or a few months of a relationship, we find that we are still unfulfilled, we blame our partners and look up with somebody more promising. This can go on and on, until we admit that while a partner can add sweet dimension to our lives, we, each of us, are responsible for our own fulfillment. Nobody else can provide it for us, and to believe otherwise is to delude ourselves dangerously and to program for eventual failure in every relationship we enter.

The Human Commodity

Modern man has transformed himself into a commodity; he experiences his life energy as an investment with which he should make the highest profit, considering his position and the situation on the personality market. He is alienated from himself, from his fellow men and from nature. His main aim is profitable exchange of his skills, knowledge, and of himself, his "personality package" with others who are equally intent on a fair and profitable purchase of the package. Life has no goal except the one to move, no principle except the one of fair exchange, no satisfaction except the one to consume.

Toy World

We all know how fond small children are for toys. The toys could be any of the thousands of varieties available in market. Toy cars, toy houses, toy games, electronic toys and the list is endless.

It is a sight to see two children fight for the same toy. There may be numerous toys around, some of them a lot better than the one they are fighting over. But they fight tooth and nail and even come to blows for the possession of that particular toy. Surprisingly, half an hour later the same toy will be resting on the floor, abandoned, none of the children even refusing to acknowledge its existence. That is because the fancy has gone and the delight of possession has gone.

What was looking most desirable and worth having at any cost, has lost its magic and appears worthless.

Adults also have their own toys. Luxury cars, palatial houses, high end gadgets etc. etc. There is also high power marketing which makes you decide what is desirable and what you should crave for. It may be your dream car, heavenly house, soulful clothes, blissful jewelry, diamonds of love, wonder gadgets, smart phones or any of the millions of consumer items you are made to think are necessary for a happy life. Just like children, people also put in their utmost efforts and a struggle of a life time to acquire these dream merchandise.

People also like to brag about their possessions as it gives them afeeling of 'I am different from you' or 'I am better than you.' They compare cars which accelerate from 0 to 100 kilometers in 5 seconds and the other which does it in 8 seconds. When we see people wasting a life time on trivial work or wasting hours in such useless talk, we wonder what they will do with the few seconds gained by way of acceleration.

Once these dream items come under one's possession, there is initial excitement and euphoria. But after some time, the excitement flattens out. The mundaneness of these material objects bursts through the reality. By the time you are done with it and go down to your routine depression, the market is ready to drive you to the next most desirable object and to make you work for it. Now you have a 'purpose' of life, something to think about, something to plan and occupy yourself with and you feel some comfort for the time being.

It takes a little deep thinking to shift from the toy world mentality to the world of reality.

Love and Knowledge

To go deeper into the aspect of knowledge, let us see the instance when we encounter a person who is angry. The commonest response is to counter him with more anger. But, when we make sincere efforts to know him better, we understand that his anger is only the manifestation of something deeper. We see him as anxious and embarrassed, that is, as the suffering person, rather than as the angry one.

Knowledge serves one more, and a more fundamental motive for love – that is to know the "secret of man." While life in its merely biological aspects is a miracle and a secret, man in his human aspects is an unfathomable secret to himself—and to his fellow man. We know ourselves, and yet even with all the efforts we may make, we do not know ourselves. We know our fellow man, and yet we do not know him. The further we reach into the depth of our being, or someone else's being, the more the goal of knowledge eludes us. Yet we cannot help desiring to penetrate into the secret of man's soul, into the innermost nucleus which is "he."

As we go deeper and deeper into the psyche of the humans, we understand that the differences in talents, intelligence and knowledge are negligible in comparison with the identity of the human core common to all men. If we perceive in another person mainly the surface, we perceive mainly the differences, that which separates us. If we penetrate to the core, we perceive our identity, the fact of our brotherhood.

The more you understand, the more you love; the more you love, the more you understand. They are two sides of one reality. The mind of love and the mind of understanding are the same.

Love Starts with the Self

We have already seen in the chapter 'Types of Love,' both ancient Greeks and the Indian wise men have proclaimed that one of the highest forms of love is 'self-love.' This is not the self as we usually think of it, but the essential self, the self that exists at the center of all of us.

What this means in practice is that we see ourselves in others and see others in ourselves.

This you can achieve only when you have shed your garbage of greed, possessiveness, jealousy and hatred.

The famous Sufi scholar Rumi has said 'Your task is not to seek for love, but merely to seek and find all the barriers within yourself that you have built against it.'

This leads us to the ultimate and highest form of love, which is compassion, when you are in love with each and every animate and inanimate objects of this universe.

This will make you what you are – a simple human being, a manifestation of the divine and an integral part of the cosmos.

This love is the breath of the Holy Spirit inspired into the human spirit and a source of spiritual fulfillment.

XIV. Unity of Being

There is only one substance and that substance we can conceive of as either Nature or God

– Benedict de Spinoza

From the 'Big Bang' episode to the present day where we see people banging each other every day, it is a journey of billions of years.

Do we have anything to learn from this journey?

Quite a lot, and it can make this world a better place to live for every one of us.

It makes us understand better what Albert Einstein has said:

"A human being is part of a whole, called by us the Universe, a part limited in time and space. He experiences himself, his thoughts and feelings, as something separated from the rest, a kind of optical delusion of his consciousness. This delusion is a kind of prison for us, restricting us to our personal desires and to affection for a few persons nearest us.

Our task must be to free ourselves from this prison by widening our circles of compassion to embrace all living creatures and the whole of nature in its beauty."

When we understand the fundamental nature of ourselves and that at core of existence all human beings are same, our differences disappear and peace and harmony prevails.

Narrow religious differences and animosities have no ground to thrive.

Again Einstein has some tips for us:

'The religion of the future will be a cosmic religion. It should transcend a personal God and avoid dogmas and theology. Covering both natural and spiritual, it should be based on a religious sense arising from the experience of all things, natural and spiritual and a meaningful unity. Buddhism answers this description. If there is any religion that would cope with modern scientific needs, it would be Buddhism.'

The English metaphysical poet and cleric John Donne, wrote in 1624.

'No man is an *Island*, entire of itself; every man is a piece of the *Continent*, a part of the *main*; ....any man's *death* diminishes *me*, because I am involved in *Mankind*; And therefore never send to know for whom the *bell* tolls; It tolls for *thee*.'

Philosophers too had been busy in this area, seeking unity in the disarrayed world they live in. Benedict de Spinoza 1632–1677, a force to reckon with in modern philosophy says: 'There is only one substance and that substance we can conceive of as either Nature or God'

When AdiShankara of India formulated his Advaita school of philosophy in eighth century, one thing which stood out is his unity of all beings. According to AdiShankara, the one unchanging entity (Brahman) alone is real, while changing entities do not have absolute existence. Any person who dives into the depths of his philosophy can never describe humanity on divisive terms or disseminate hatred among them.

The same ideas, the Sufi philosopher Rumi has expressed eloquently:

> *I, you, he, she, we—*
> *in the garden of mystic lovers,*
> *these are not true distinctions.*

Only love can show us how to live in harmony with nature and with each other.

Love is the binding force that unifies Creation on all levels. It brings together all forms of life in a sacred Whole, including those forms that are conscious of this greater Unity, and those that are less conscious.

Human being, as part of a whole, always strives for something more. Though what we strive for or how we drive at it may be different, confusing or even incomprehensible, the striving is always there.

Thomas Carlyle has put it beautifully:

"The misfortune of man has its source in his greatness. For there is something infinite in him and he cannot succeed in burying himself completely in the finite."

This beautifully explains the thirst every human being has for 'something more' – the endless urge to be better and more complete.

Every person is like a potent healthy seed. He carries the core of massive blossoms inside. Whether he shrivels and terminates inside his shell or breaks his shell and blossoms into glorious existence is his choice to make.

Every person, at his core is wholesome, infinite and blissful.

As there are very few people who have realized this, obviously there are road blocks on our way forward.

What ties us down, from realizing our real selves are the loads of garbage we carry.

To start the inventory of the garbage, the foremost may be the worldly identities thrust on us from birth – like our name, religion, nation, race, pedigree etc. etc. They get ingrained in our system and form the greatest obstacle to the unity of mankind.

Apart from the general garbage we carry, there is a stockpile of personal junk – like ego, greed, hate, jealousy, possessiveness etc. etc.

It is possible to realize self only when a person is courageous enough to break his shell and unload all his garbage.

XV. The Homosapien – A Natural History

"The marks humans leave are too often scars."

– John Green, The Fault in Our Stars

So far, we had a glimpse of what we are and from where we could have come!

It is time we also examine our immediate roots and where we are going and what our destiny is likely to be.

The species Homosapien is just one species out of a total of 1.3 million species that have been identified and described as of now. But the truth is, many more live on Earth. The most accurate census, conducted by the Hawaii's University, estimates that a total of 8.7 million species live on the planet.

It implies we still have not examined or classified 86% of the terrestrial species and 91% of the marine species.

Still more bewildering is the fact that more than 99 percent of all species, amounting to over five billion species, that ever lived on Earth are estimated to have died out.

In the light of above facts, we certainly have an obligation to examine what our fate is likely to be.

It will be useful to know what we were before and how we came to be what we are now, so that we can speculate on what we will be in future.

We are the latest and most recent inhabitants of this earth. To put things in perspective, bacteria had been living on this earth for 4,000,000,000 years. Cockroaches were swarming the earth 320,000,000 years ago and scorpions originated 430,000,000 years ago. Compared to them, we are recent, modern humans making their appearance only 200,000 to 100,000 years ago.

First, we will have a small study of our ancestors who brought us here.

Our story starts with the Australopithecus, which literally means 'Southern Ape,' and a significant player in human evolution.

The genus Australopithecus apparently evolved in eastern Africa around 4 million years ago before spreading throughout the continent. They became extinct two million years ago.

The genus Homo evolved from Australopithecus at some time three million years ago.

At this point, it is better to be clear about what a genus or species stand for.

A genus is a group of animals or plants that includes several closely related species. A genus is a larger group and the different species it contains do not normally interbreed.

A species is a group of plants or animals whose members have the same main characteristics and are able to breed with each other to produce off springs.

Around two million years ago, one of the australopith species evolved into the genus Homo of which Homo Habilis is an example. This eventually gave rise to modern humans, H. sapiens sapiens.

Homo habilis is an intermediate species of Homo, between Australopithecus and the somewhat younger Homo erectus. It lived between roughly 2.1 and 1.5 million years ago and became extinct.

Homo erectus is a fellow member of our own genus. Alive from 1.89 million years ago to 143,000 years ago, he was the first to leave Africa and the first to master the use of fire. H. erectus is known in Africa as Turkana Boy, in China as Peking Man, in Indonesia as Java Man and in Europe as Tautavel Man.

Homo erectus had shorter arms, which were a departure from Homo habilis and Australopithecus, indicating that tree-climbing ability had finally been lost. The longer legs were better suited to running and walking long distances – a trait that doubtless helped Homo erectus migrate into Asia. Ranging in height from 4 foot 9 inches to 6 foot 1 inch and weighing 88 to 150 pounds, Homo erectus was the first ancestor to approximate modern humans in size.

H. erectus eventually became extinct throughout its range in Africa, Europe and Asia by about 143,000 years ago.

A particular population H.erectus continued to evolve. That population gave rise to Homo heidelbergensis, which then gave rise to Homo sapiens and Homo neanderthalensis species separately.

Homo heidelbergensis is an extinct species or subspecies of archaic humans in the genus Homo. They lived from about 700,000 to 300,000 years ago, and the fossils of these extinct species have been found in Southern Africa, East Africa and Europe.

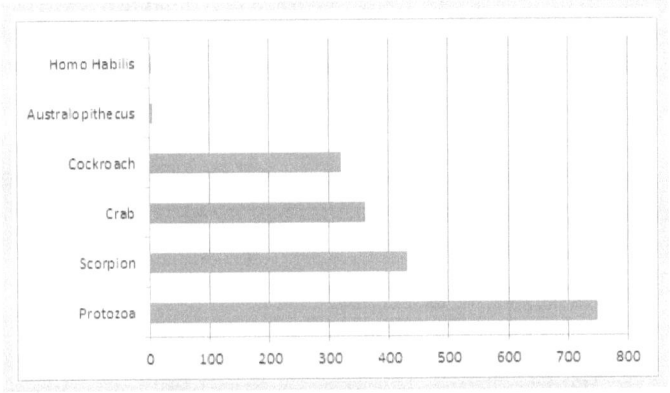

26. Period of existence of Hominids, compared to other species – in Millions of years.

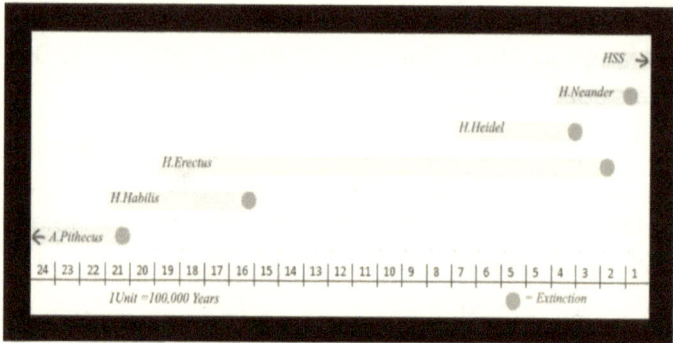

27. Chart shows duration of existence of various Homo species on earth.

You can see that we are the only surviving and continuing species of the genus Homo, while all our brethren have become extinct. It also shows we have occupied the earth for the shortest period of time (so far), compared to any other of our related species. Each division represents 100,000 years ago and dot at end denotes extinction.

Neanderthals are a species or subspecies of archaic humans of the genus Homo. They lived within Eurasia from about 400,000 years to until 40,000 years ago and became extinct.

According to genetic and fossil evidence, older versions of Homo sapiens evolved only in Africa, between 200,000 and 100,000 years ago, with members of one branch leaving Africa by 90,000 years ago. Over time, they replaced earlier human populations such as Neanderthals and Homo erectus throughout the world.

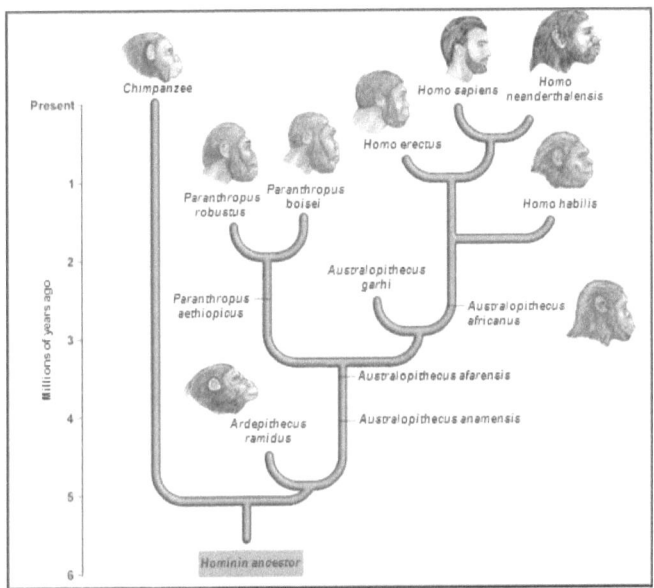

28. Hominid Evolution Tree.

The above narration is about anatomically modern humans. We do not know whether they behaved similar to recent or existing humans. Behavioral modernity is taken to include fully developed language (requiring the capacity for abstract thought), artistic expression, early forms of religious behavior, increased cooperation and the formation of early settlements. Behavior modernity also includes the production of articulated tools from lithic cores, bone or antler. This change seems to begin around 50,000 to 40,000 years ago, and also coincides with the disappearance of archaic humans such as the Neanderthals.

So, the age of behaviorally modern homosapiens seems to be just about 50,000 years.

Inter Breeding

Some recent research on DNA profile of fossil specimen and discovery of new fossils have made the evolution story more complex and interesting.

DNA evidence shows that human evolution should not be seen as a simple linear or branched progression, but a mix of related species which show hybridization. Hybridization is the process by which members of two different species mate and produce an entirely new offspring which has genetic material of both the parents.

In fact, inhuman evolution, genomic research has shown that hybridization between substantially diverged lineages is the rule, not the exception. Furthermore, it is argued that hybridization was an essential driving force in the emergence of modern humans.

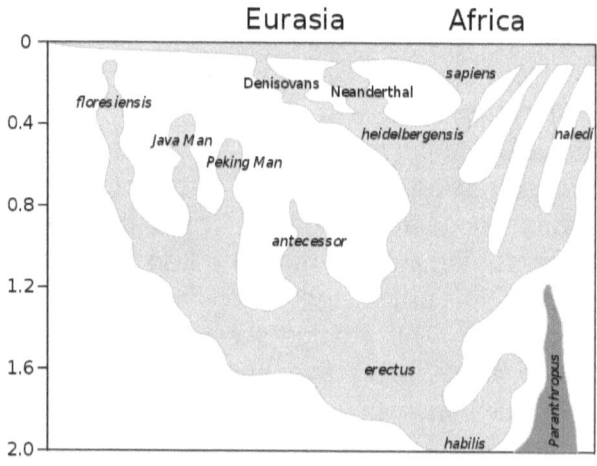

29. The Evolution, Branching and Merging of the Genus Homo.

The vertical axis shows the duration in millions of years.

You can note that by about 100,000 years ago, various species of Homo have started merging together. At point '0,' that is present time, there is only one species, Homo sapiens sapiens. All other species have become either extinct or merged into us.

About 20% of the Neanderthal genome has been found incorporated in the modern human population. This was arrived at after extensive analysis of East Asian and European genomes. This seems to substantiate the concept that there was interbreeding between the Neanderthals and the modern humans.

Yet another find in Siberia recently has taken the narrative to still more complex outcomes.

In 2010, scientists found an undated finger bone fragment of a juvenile female found in the Denisova Cave in the Altai Mountains in Siberia. This was the discovery of the Denisovans or Denisovahominins which are an extinct species or subspecies of archaic humans of the genus Homo. The nuclear genome from this specimen suggested that Denisovans shared a common origin with Neanderthals and that they ranged from Siberia to Southeast Asia. Denisovans also lived among the ancestors of some modern humansand interbred with them.

A comparison with the genome of a Neanderthal from the Denisova cave revealed local interbreeding, with local Neanderthal DNA representing 17 percent of the Denisovan genome. They also revealed evidence of interbreeding with an as yet unidentified ancient human lineage.

136 ♦ *Time Space and You*

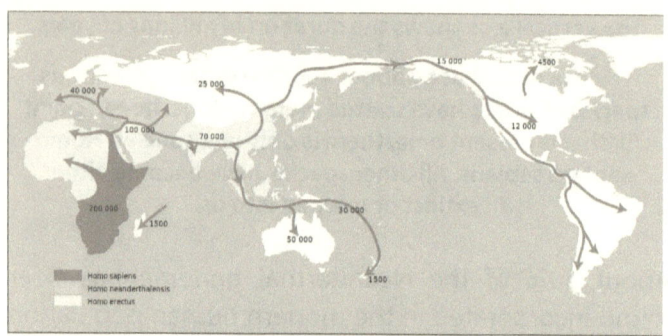

30. Spread of human beings around the world.

Most studies attest to the fact that the modern humans made their appearance in Africa about 200,000 years ago. They moved out of Africa about 100,000 years ago to gradually occupy the whole world. The numbers show how many years ago the movement into different geographical locations took place.

Several types of humans, including Denisovans, Neanderthals and related hybrids, may have dwelt in the Denisova Cave in Siberia over thousands of years. But it is unclear whether they ever cohabited in the cave.

Scientists opine that Denisovans may have interbred with modern humans in New Guinea as recently as 15,000 years ago.

Where are we Going?

Taking into account, the geological and biological history of our planet earth, we can call ourselves new arrivals to this place. We occupy only a fraction of percentage of duration of life on earth. The story of our ancestors is also not encouraging. Almost all of them have become extinct, after making a mark in this world for some time. Our existence has been so short, we do not have much to speculate. If we base our speculations on our ancestors, the inferences are not comforting.

In near future, will our scientists with their incessant tampering with human genes produce so called 'Super Humans'? If they do so, how will the 'Super Human' look like and what will be his attributes?

Our concerns still further in time will be, is it possible we may evolve in to a different species? Powerful forces of nature which created and molded us are always at play and we do not know how we will be shaped.

Will we branch out and give rise to more species?

Is it possible we will interbreed with differentiated species and give rise to radically different species?

Do human beings will be fortunate (or unfortunate) enough to survive that long to witness such events?

Alfred Lord Tennyson, in his famous poem 'The Brook' has said

'For men may come and men may go,

But I go on forever.'

It makes us wonder whether really men will keep coming and going in this world and what type of men it will be.

Even if we assume our natural history allows us to 'go on forever,' we have surprises in store for the near and distant future.

XVI. Unity of Doom

*"Everyone, deep in their hearts, is waiting for
the end of the world to come."*

– Haruki Murakami, 1Q84

Having seen the origin of life and the possible course of the humans, the last and natural question arises – How we may end, how the world we live in may end and if so, will there be any pattern to it?

Death is the common fate shared by every living organism – in fact a very nature of life. In about 100 years from now, every human being who is alive now is likely to be dead. Still, we humans are always curious to know how the world will be in the short term, that is some hundreds of years from now, as well as some millions of years later.

To be honest, these conjectures do not paint very rosy pictures for the human kind. Nevertheless, we will have a peep into them.

Eschatology

Eschatology is the term which refers to the question of end of life, end of the world and end of everything. This 'Dooms day' type of belief occupies people's mind across several religions and several regions. It is mysterious why this idea is so prevalent! Is it possibly connected to the subconscious fear of death most people harbor? We cannot be sure. May be it is more convenient to think the world will come to an end and everybody will perish rather than to think that one can die any day and that death is an absolute certainty.

There are extensive speculations over the end of the world based on religion, mythology, folklore and science regarding this. We can call them speculations because we have not experienced the end, as we conceive it. Though scientific conjectures try to base their statements on available facts and scientific extrapolation of facts, the other eschatological observations do not seem to bother about any such basis.

We will have a look in to these stories and also what science has to tell about them.

Mythology

The Abrahamic faiths, namely the three major religions of Judaism, Christianity and Islam depict end of time scenarios containing themes of transformation and redemption.

In Judaism, the term "end of days" makes reference to the Messianic Age and includes a coming together of the exiled Jewish diaspora, the coming of the Messiah, the resurrection of the righteous, and the world to come

Armilus is an anti-messiah figure in medieval Jewish eschatology, comparable to medieval interpretations of the Christian Antichrist and Islamic Dajjal. He will conquer whole Earth and will centralize in Jerusalem and persecute the believers until his final defeat at the hands of Messenger of God or the true Messiah. His inevitable destruction symbolizes the ultimate victory of good over evil in the Messianic Age.

Christianity depicts the end time as a period of tribulation that precedes the second coming of Christ, who will face the Antichrist along with his power structure and usher in the Kingdom of God. Antichrist is also referred by the term pseudokhristos or "False Christs" in the Gospels.

Some of the major components of Christian eschatology are death and the afterlife, Heaven and Hell, the second coming of Jesus, the resurrection of the dead, the rapture, the tribulation, millennialism, the end of the world, the Last Judgment, and the New Heaven and New Earth in the world to come.

In Islam, the Day of Judgement is preceded by the appearance of the al-Masih al-Dajjal, and followed by the descending of Isa (Jesus). Isa will triumph over the false messiah, or the Antichrist, which will lead to a sequence of events that will end with the sun rising from the west and the beginning of the Qiyamah (Judgment day).

Al-Masih ad-Dajjal (which in Arabic means "the false messiah, liar, the deceiver") is an evil figure in Islamic eschatology. He is to appear, pretending to be al-Masih (i.e. the Messiah), before the Day of Resurrection. He is an anti-messianic figure, comparable to the Antichrist in Christian eschatology and to Armilus in medieval Jewish eschatology.

In Abrahamic religions, a common theme is the Messianic Age, where in at a future period of time on Earth a messiah or a messenger of God will reign and bring universal peace and brotherhood, without any evil. An anti God or evil figure, eventual destruction of the world and resurrection and revival also form essential components of their dooms day.

Non-Abrahamic faiths tend to have more cyclical world-views, with end-time eschatologies characterized by decay, redemption, and rebirth.

In Hinduism, the end time occurs when Kalki, the final incarnation of Vishnu, descends atop a white horse and brings an end to the current Kali Yuga.

Kalki will amass an army to "establish righteousness upon the earth" and leave "the minds of the people as pure as crystal." Those left, transformed by virtue, will be the new seeds for a higher form of humanity, and humanity will begin again

In Buddhism, the Buddha predicted that his teachings would be forgotten after 5,000 years, followed by turmoil. A bodhisattva named Maitreya will appear and rediscover the teaching of dharma. The ultimate destruction of the world will then come through seven suns.

In the "Sermon of the Seven Suns" in the Pali Canon, the Buddha describes the ultimate fate of the world in an apocalypse that will be characterized by the consequent appearance of seven suns in the sky, each causing progressive ruin till the Earth is destroyed.

In a series of changes as each sun appears, gradually all plants and vegetation burn out, rivers and oceans evaporate and the Pali canon further states,

'Again after a vast period of time a sixth sun will appear, and it will bake the Earth even as a pot is baked by a potter. All the mountains will reek and send up clouds of smoke. After another great interval a seventh sun will appear and the Earth will blaze with fire until it becomes one mass of flame. The mountains will be consumed, a spark will be carried on the wind and go to the worlds of God… Thus, monks, all things will burn, perish and exist no more except those who have seen the path.'

The sermon completes with the planet engulfed by a vast inferno. The Pali Canon does not indicate when this will happen relative to Maitreya.

In Egyptian mythology, end is described in a passage in the Coffin Texts and a more explicit one in the Book of the Dead. Atum, the creator of God, says that he will one day dissolve the ordered world and return to his primeval, inert state within the waters of chaos. All things other than the creator will cease to exist, except Osiris, who will survive along with him

In Norse mythology, Ragnarök is a series of events, including a great battle, predicted to lead to the death of a number of great figures (including the Gods Odin, Thor, Týr, Freyr, Heimdallr and Loki), natural disasters and the submersion of the world in water. After these events, the world will resurface anew and fertile, the surviving and returning gods will meet and the world will be repopulated by two human survivors.

Eschatological beliefs are so extensive and worldwide, going into their details will be beyond the scope of this book.

Engineering a Doom

'Human extinction' is the hypothetical end of the human species and is a subject of active debate in many fora. This may result from natural causes or it may be the result of human action.

The likelihood of human extinction in the near future by wholly natural scenarios, such as a meteorite impact or large-scale volcanism, is generally considered to be extremely low.

The biggest threat to human survival is human beings themselves.

One of the major crises our species face is loss of habitat and Stephen Hawking said at an event in 2016 "Our earth is becoming too small for us, global population is increasing at an alarming rate and we are in danger of self-destructing… I would not be optimistic about the long-term outlook for our species."

There is no other population of large vertebrate animals like humans in the history of the planet that has grown so much, so fast, or with such devastating consequences to fellow earthlings. The growth of population occurred rapidly from 1 billion in 1800 to over 7 billion today. With the population going up, people will annually absorb more and more primary productivity of the Earth. This may eventually lead to a crash when the population grows beyond the capacity of its environmental sustainment and reduces its capacity below the original level

The most dreadful scenario is where human extinction is brought about by the humans themselves. For such anthropogenic extinction, many possible scenarios have been proposed: human global nuclear annihilation, biological warfare or the release of a pandemic-causing agent, overpopulation, ecological collapse, and climate change. We will consider some of them.

Nuclear Annihilation

The world right now is supposed to have a stockpile of more than a thousand nuclear bombs. Since it involves a great deal of secrecy, it is quite difficult to be precise about their number or capability. But one thing we can be sure about is that the world has an unnecessarily large number and unimaginably powerful nuclear devices.

On 30th October 1961, the erstwhile Soviet Union exploded the RDS-220 hydrogen bomb, popularly known as the "Tsar Bomba." It is the biggest and most powerful thermonuclear bomb ever detonated. Its explosive power is equivalent to simultaneous detonation of 3,800 Hiroshima bombs. If we consider the great physical damage and loss of life Hiroshima incurred the instant bomb was dropped, and the fact that the people there still suffer from the radiation induced long term ill effects, Tsar Bomba looks like a model of the dooms day in the making.

Americans are not far behind in the development of such lethal weapons and are stockpiling them. Some other countries have also followed suit. All it shows is that we have enough and more of these lethal weapons to destroy ourselves.

Imagine a situation where there is a nuclear warfare, by choice or by even accident (as most of these draconian devices are highly automated and are operated by finger touch buttons), we have a very good recipe for the end of humanity.

Our Quarrel with Virus and Bacteria

The smallest organism which inhabits the earth is virus. In fact they are so small, they cannot be seen even by ordinary light microscopes and one needs electron microscope to visualize them. They usually measure less than 200 nanometers.

Viruses are entirely dependent a host cells for metabolism and for reproduction, making them obligate intracellular parasites

Virus infections are very difficult to treat as they get ingrained into the protein structures of the body. So, no antibiotic works on them. The common mode of terminating viral infections in humans is by developing immunity against them. The best way to prevent the viral infection is to get vaccinated against each one of them. But the viruses also mutate and develop new strains, so that it becomes extremely difficult for people to fight against them.

Growing number of disease producing bacteria are becoming harder, and sometimes impossible to control. Antibiotics, our killer weapon for bacteria, become less effective over time. The bacteria seem to learn how to become resistant to our antimicrobial agents, paying no heed to the billions of dollars put into research to develop those drugs. These smart operatives use several tactics to hoodwink the antibiotics. They can do enzymatic degradation of antibacterial drugs, alter the bacterial proteins that are targeted by the antimicrobial agents or change the permeability of their membrane coats so that antibiotics cannot enter inside them.

Bacteria can also undergo mutation or alteration of their DNA to counter the antibiotics. Such changes are even passed on to their progeny.

Though every now and then we come out with miracle drugs to eliminate the disease producing bacteria, after a while they no longer perform miracles. It is the bacteria which perform miracles on them.

"Superbugs" is a term used to describe strains of bacteria that are resistant to the majority of antibiotics commonly used today. Resistant bacteria that cause pneumonia, urinary tract infections and skin infections are just a few of the dangers we now face. Centers for Disease Control and Prevention, USA states that several thousand deaths occur each year due to the antibiotic resistant organisms.

Our battle with the pathogenic bacteria has been an ongoing process for a long time. Indiscriminate use of antibiotics, unhygienic habits and habitats will be greatly detrimental to the humans with serious repercussions.

Epidemics of virulent viral infections still threaten human survival worldwide. A pandemic of SARS viral infection (2002–2003) affected 26 countries and resulted in more than 8000 cases. It also killed more than 900 people.

Now, at the time of writing this book, coronavirus (Covid 19) infection is at its extremely dangerous phase. It has become a pandemic, affecting 213 countries. It has already infected more than six million and killed about three hundred and sixty thousand human beings. Billions of people arein quarantine, hiding from the virus in homes or health centers.

An even more deadly scenario is the use of these microorganisms in biological warfare by our fellow humans. They are designed to be lethal, resistant and not easily controllable. If these pathogens go out of control, by design or by accident, we will be staring at a catastrophe.

A classical or rather horrific example of the power of the bacteria is Clostridium botulinum. It is a rod-shaped, anaerobic motile bacterium which produces the very potent neurotoxin botulinum. Scientists have evaluated that just one teaspoon of the organism, spread properly over our dear world, is capable of wiping out the human race. It is said just one kilogram of the toxin extracted from this bacteria is enough to kill every person in this world.

It is another irreconcilable fact that these same botulinun toxins are used in beauty treatment for removing wrinkles in the so called 'Botox Treatment.'

No wonder people say beauty can be lethal!

Futuristic Human Technologies

New emerging technologies could bring about new extinction scenarios, such as advanced artificial intelligence, biotechnology, or self-replicating intelligent robots. These are vast subjects by themselves.

Mother Earth

The natural history of earth does show some possible scenarios which may be intimidating to the human kind. One such is the supercontinent cycle, which is the quasi-periodic aggregation and dispersal of Earth's continental crust.

We travel several millions of kilometers per hour through the space, perched on our planet, without being aware of it. Now, we come to another serious journey, where we travel one or two inches per year on the surface of the earth.

It all started with the theory of continental drift proposed by the scientist Alfred Wegener. In 1915, Wegener published a paper explaining his theory that the continental landmasses were "drifting" across the Earth, sometimes plowing through oceans and into each other. He called this movement continental drift. The scientists of that day found it difficult to believe him.

Wegener proposed that all of Earth's continents were once part of an enormous, single landmass called Pangaea.

Wegener produced several evidences from biology, botany, and geology to substantiate his theory of Pangaea and continental drift.

As major biological evidence, he took the fossil record of the ancient reptile mesosaurus which lived 299 million to 271 million years ago. These fossils are found only in southern Africa and South America. Mesosaurus, which means "middle lizard, is only one meter (3.3 feet) long. It is a freshwater reptile and it could not have swum several thousand kilometers across the salty Atlantic Ocean. The presence of such mesosaurus fossils suggested that a single habitat with many fresh water lakes and rivers existed millions of years ago. This, in practical terms means Africa and South America should have been one single land mass millions of years ago.

Wegener produced evidence additionally from plant fossils from the frigid Arctic archipelago of Svalbard, Norway. These fossils were of tropical plants, which are adapted to a much warmer, more humid environment. In no way they could have grown in the freezing Arctic. The presence of these fossils suggests that Svalbard, now in the Arctic, once had a tropical climate and had moved from a place of tropical climate to the present one.

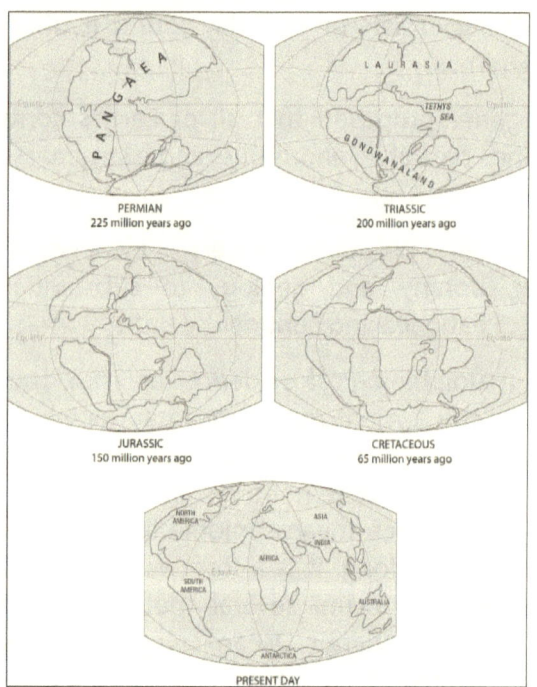

31. Pangea – The Super Continent, as it was 225 million years ago and its gradual dispersal movement to the present day.

Finally, Wegener used stratigraphyto prove his point. It is a branch of geology concerned with the order and relative position of different layers of earth's crust and their relationship to the geological timescale,. The east coast of South America and the west coast of Africa seem to fit together like pieces of a jigsaw puzzle, and Wegener discovered their rock layers "fit" just as clearly. South America and Africa were not the only continents with similar geology. Wegener discovered that the Appalachian Mountains of the eastern United States, for instance, were geologically related to the Caledonian Mountains of Scotland of Europe.

Scientists of today have come to the conclusion that Pangaea did exist about 240 million years ago. By about 200 million years ago, this supercontinent began breaking up. Over millions of years, Pangaea separated into pieces that moved away from one another. These pieces slowly assumed their positions as the continents we recognize today.

Scientists also understand today that the continental movement is due to the movement of the tectonic plates on the crust of the earth, which keep moving slowly all the time. All of the Earth's continents float on tectonic plates, which glide slowly over a plastic-like layer of the upper mantle.

This brings us to the Supercontinent cycle, which is the quasi-periodic aggregation and dispersal of Earth's continental crust. It constantly modifies Earth's surface. The tectonic plates which support the surface of the earth periodically come together to form one big land mass called a 'Super Continent' and also disperse again. One such complete supercontinent cycle is said to take 300 to 500 million years.

Scientists think that several supercontinents like Pangaea have formed and broken up over the course of the Earth's lifespan. These include Pannotia, which formed about 600 million years ago, and Rodinia, which existed more than a billion years ago.

The most recent supercontinent, Pangaea, formed about 300 million years ago.

The breakup of Pangaea started with the opening of the Atlantic Ocean, which split North America from Asia and South America from Africa. India emerged from where it had lain wedged between Africa, Antartica and Australia for about 500 million years. India drifted northwards across the Southern Ocean and smashed into Asia to make the Himalayas. Australia separated from Antarctica and headed off for South-East Asia. Africa moved north towards Europe.

The continents are still moving today. Himalayas keep gaining height at the rate of between 1 cm and 6.1 cm each year.

The North American and Eurasian tectonic plates are separated by the Mid-Atlantic Ridge. The two continents are moving away from each other at the rate of about 2.5 centimeters (1 inch) per year.

Australia sits on the fastest moving continental tectonic plate in the world. Australia is moving north by about 7 centimetres each year, colliding with the Pacific Plate, which is moving west by about 11 centimetres each year. Australia has shifted by 4.9 feet since the last adjustment was made to GPS coordinates in 1994.

One of the definitive predictions on future is that Africa is likely to continue its northern migration, swallowing the Mediterranean sea and driving up a Himalayan-scale mountain range in southern Europe by pushing against it. Australia is also likely to move north and merge with the Eurasian continent.

Scientists are of the opinion we are headed for a new supercontinent in about 250 million years from now. Several names have been already proposed for the next super continent, which include Pangaea Ultima, Pangaea Proxima, Neopangaea, Amasia and Pangaea II.

The outward telltale evidence of such movements is occurrence of earthquakes and sudden eruption of volcanoes. They are bound to continue as the tectonic plates keep moving.

As humans, where do we stand!

We stand nowhere.

The natural history of earth may be intimidating to the human kind. When geology rebuilds the lost worlds and redesigns the continents it does not care where you stand or what you do.

What you hold on to as your street, your town or your country is a result of the past movements of the earth. The future movements, if you are alive that long, may not give you the same location or address.

Father Sun

Another Dooms Day scenario involves our solar system. As years go by, the luminosity of the Sun will steadily increase, resulting in a rise in the solar radiation reaching the Earth. This will cause a decrease in the level of carbon dioxide in atmosphere due to weathering of silicate minerals which absorb quite a lot of this gas. In about 600 million years from now, the level of carbon dioxide will fall below the level needed to sustain carbon fixation photosynthesis used by trees. The long-term trend is for plant life to die off altogether. The extinction of plants will be the demise of almost all animal life, since plants are the base of the food chain on Earth.

In around 1 billion years from now, the Sun's brightness may increase as a result of a shortage of hydrogen. Ensuing heating of the outer layers of earth may cause its oceans to evaporate, leaving only minor forms of life.

In about 7 to 8 billion years from now, the Sun is predicted to become a red giant and the Earth will probably be engulfed by an expanding Sun and destroyed.

So What?

Is there anything to be happy about in such gloomy scenarios?

We do have some silver lining to cling on to.

Though our past is not very colorful and future has nothing much to cheer, we are neither the 'Past' nor the 'Future.'

We are the 'Present.'

Right now we are living in a very beautiful world and enjoying the miracle called life bestowed on us.

Even though we are not sure what will happen in a thousand years from now, one thing we can be absolutely sure. You and I will be not be around, at least not in the present form, to witness that time andwill be long gone by then.

It gives us more reasons to enjoy our blessing called Life and adore every moment of it.

XVII. Can God Save Us?

With God, all things are possible ....

'Security Blanket' of Religion

Imagine you are on flight on a long awaited tour you had been planning for some time. You are 40,000 feet above the Pacific Ocean. Suddenly you hear the announcement that the plane seems to have developed some snags, you may have a rough ride and to kindly fasten your seat belts.

Try to visualize the fear and gloom inside the flight.

Almost all the three hundred plus passengers inside the plane start praying, irrespective of the religion and irrespective of the personal doubts some people may have about the God. Some of the faithful feel God will answer their prayers and most feel they are doing what best they can do at this juncture.

This is a familiar situation where fervent religious sentiments take over humans. It happens all the time in modern situations like bombings, earth quakes or other general or personal disastrous situations.

This had been happening from very ancient times.

Only that they prayed before going to war, start hunting or sail out into the sea or when nature struck them with terrible fury.

32. Purification Rites.

**Natives of Papua New Guinea, an island in the South Pacific Ocean purifying themselves by sprinkling water on themselves before a religious ritual.
(Photo by Author, 2012)**

33. Artifacts at the Papua New Guinean rituals.

Religion has survived and thrived for more than 100,000 years, or shall we say from the time humans became capable of abstract thinking. It exists in every culture, with more than 85 percent of the world's population embracing some sort of religious belief.

There is a 'security blanket concept of religion' which has many proponents. It explains why people pray during a crisis, and why people living in the most miserable places on earth are universally religious.

Religious ritual is a source of comfort when people are distressed. Recent research shows that just as in modern psychotherapy, they can provide relaxation and meditation to the distressed soul. It may fill the human need for finding a meaning in life, sparing us from existential anxiety.

34. Some people believe God can have intermediaries.

A Hindu religious Sadhu blessing the devotees from a cave in Uttarkhand, India. (A photo by Author, 2014)

In the ancient times, life was always full of difficulties. That is why religion is a human universal. Instead of getting drowned in personal misery, self-transcendence or a feeling of otherworldliness gave them solace and something to go by.

On the other hand, in societies that experience a good quality of life, religion loses its importance, and atheism breaks out.

In prosperous, socially secure countries with very high living standards, the people tend to be atheistic. In countries like Sweden, Norway and Iceland, more than 80% of the people do not have belief in a conventional god or religion.

China has the world's maximum number of convinced atheists, estimated to be more than 200 million and a very high percentage of non-religious people. It has several reasons going for it.

The Communist Party of China, which is ruling the country since 1949, regards religion as a means of oppressing the working class of people. All religious movements were severely suppressed under Mao Zedong throughout his 27 year reign until 1976.

Another major factor is the influence of Confucianism, the very ancient philosophical doctrine followed by most Chinese. Confucianism is notable for its lack of a belief in a supernatural deity.

Demographic analysis show people in some of least prosperous and socially distressed countries have the most religious persons. In Ethiopia, Somalia and Bangladesh, more than 99% of the people consider themselves religious.

Religion as Social Organization

Religion may serve another key purpose—it allows people to live in large, cooperative societies. In fact, the use of religion as a social tool, a basic platform for large social organizations, may largely explain its staying power and cross-cultural ubiquity.

'God Spot' in Brain

Religious beliefs and rituals are found in every society studied by anthropologists. This implies that religious or spiritual experience is a universal characteristic of human beings. An evolutionary perspective on religion implies that humans are inherently susceptible to religious views.

It has been found that spiritual experiences (including religious experiences) have a neural basis. This led to a popular postulate that there is a 'God Center' in the brain. Although there is no single "God spot" in the brain, feelings of self-transcendence are associated with reduced electrical activity in the right parietal lobe, a structure located in the brain above the right ear.

New research from University of Missouri has indicated that there is no one God-spot in the human brain and that in fact spirituality induces stimulation across all lobes in the brain. This study also sees no difference in the brain response of a Buddhist or Christian when religious experiences occur.

The findings also claim that even the mind of die-hard atheists is stimulated in the same way as a Christian or Muslim when they experience their version of spirituality.

35. **Offerings of incense sticks in a Buddhist temple.**

> The incense stick burns itself completely into ashes and yet fills the air with a pleasant smell. This ritual basically denotes human virtue of sacrificing oneself for society.

To sum up, modern research seems to indicate that there are all kinds of spiritual experiences that Christians might call closeness to God, people of different cultures and faith may call transcendental, divine or 'other worldly' experiences and atheists might call an awareness of themselves – all produce the same neural type of neural response in the brain.

Can we assume the brain is secular in its response to various religious sentiments!

Religion and Spirituality

Religiosity and spirituality are not mutually contradictory as the boundary can be very vague for most people. We might say that spirituality involves some kind of "search for the sacred," and religion is the readily available, group-validated and organized means and methods of this searching.

One who has strong religious beliefs has no need to test one's intellectual or spiritual capabilities or explore new horizons for his personal quest. His passage becomes simple, easy and non-controversial with ready-made tenets to guide him.

Some Dissenting Notes

There had also been loud voices against religion from prehistoric times.

Karl Marx called religion as a calming "opiate of the people" and a tool in the hands of the capitalists to rule over the working class.

Sigmund Freud, the famous psychiatrist and founder of the Psychoanalytic School, has characterized religious beliefs as pathological, seeing religion as a malignant social force that encourages irrational thoughts and ritualistic behaviors.

Some psychologists, including some Christian scholars say that the doctrines of original sin and eternal damnation in Christianity can cause psychological distress. They point out that Christian scriptures say people are guilty and responsible, and face eternal punishment. Yet they have no ability to do anything about it. Psychologists propose conditions like post-traumatic stress disorder, clinical depression or anxiety disorders can be precipitated by such teachings.

To Summarize,

religion gives you ready made answers to all your metaphysical questions. It is a panacea for your fear of annihilation and existential boredom. It teaches you what to believe and how to behave. Following those norms make you an acceptable and esteemed person to the society and the wide religious group you belong. This gives you and your family social security and other advantages you crave for and makes you very comfortable within your circle.

If you want to come out of this 'comfort zone' and explore yourself as to what you are, again religion may place obstacles in your way.

Religion dissuades you from questioning the contents of the 'faith' or having alternative ideas which do not conform to the accepted tenets. Such acts are detested and are punishable. Though in modern times, you may not be called for inquisition or burnt alive for the same, the consequences may be serious in conservative societies.

Whether you have the 'free will' to follow your heart and explore new horizons or happy to plod the trodden path is a very personal choice.

XVIII. Disunity of Man

Disagreements don't cause disunity, a lack of forgiveness does.

– Loren Cunningham

From the Unity of origin to the unity of Doom, it has been a panorama of ups and downs and miraculous, unbelievable and often intimidating facts of nature.

Do we have anything to learn from them?

Yes, a plenty!

Forces of nature are unimaginably great even to understand them. Thinking of influencing them is a mad man's dream. We insignificant creatures keep riding on a tiny planet at millions of kilometers per hour, over whose speed or direction we have absolutely no control. Though we have very little say on the grandiose operations of the Universe, we are an integral part of one unity called the cosmic process.

Consciousness

We humans are unique in that we are endowed with something called 'Consciousness,' which we suppose no other living or non-living material possesses, at least in the quality and the extent to which we possess. Consciousness, literally means being aware of one's own existence, sensations, thoughts and the immediate environment. But in practice, in means much more than that and scientists and philosophers are still 'consciously' breaking their heads to find out what it is. This consciousness and abstract reasoning is the one which makes us explore the earth and the universe, devise cars and missiles and innumerable powerful gadgets and machineries which make our life comfortable or miserable, depending on how we use them.

Consciousness is also the one which makes us ask the questions 'What are we?' and 'Where are we coming from?' It cannot lie idle and is constantly at work.

We are also hopelessly trapped in incessant streams of compulsive thinking. This relentless thinking about worldly pursuits has made us lose the ability to sense the interconnectedness of all that exists. This results in loneliness becoming an inescapable facet of human existence.

Loneliness and Boredom

Loneliness, far from being a rare and curious phenomenon seems to be a central and inevitable fact of human existence. When we try to analyze the greatest of the human beings who occupied the earth or the simplest of the citizens, we find that nobody is exempt from this scourge of loneliness.

But what is loneliness, so universal and so intimidating?

Being lonely seems to be about not feeling connected in a meaningful way to others, to the world, to life.

Many people are lonely even though they have acquaintances and activities. Having hundreds or thousands of "friends" on Facebook, Whatsapp or other social networking websites is not the same as having someone to share your anxieties or intimate fears. One of the loneliest experiences may occur when you are in a crowd of people you do not feel connected with or when you are with a life partner or friend with whom you feel no bonding. It is common in our urban setting to have thousands of people live together in overcrowded apartment complexes or meet hundreds of people in a day or have a blast of a party with food and drinks and still feel extremely lonely.

Loneliness is a different experience than solitude. Solitude is being alone by choice and wanting that aloneness or being comfortable with it. Loneliness means there is a discomfort—you want to be more connected to others.

People always try to drown their loneliness in some type of activities. A person can be a workaholic, shopaholic or a plain alcoholic – the type of activity does not matter. But the ultimate aim is not to be aware of this predicament. They can also distract themselves in other types of activities like sports, music, partying, reading, travel, drug abuse etc. They can also engross themselves in umpteen forms of entertainment our consumer culture has spawned to engage our lonely souls for a while and make money out of it. These workouts we can call pleasures, which are short lived, serve to distract you from your Self for some time and then return you to your original state of anguish in no time. In no way they can be equated with the state of happiness or bliss.

36. People Busy in Pastimes.

A theme park in Dubai, where people seem to be extremely busy in going through the enjoyments.

When such distractive activities are not available or their repetitive nature makes them repulsive, the person gets in to a state of boredom, apart from being lonely.

Now, what is boredom?

Boredom is a condition characterized by perception of one's environment as dull, tedious, lacking in stimulation and accompanied by an urge to get out of such state.

Without stimulus or focus, the individual is confronted with nothingness, the meaninglessness of existence and experiences existential anxiety.

Many existential philosophers think boredom is an essential human condition and people struggle all the time to come to terms with it. It can also be the source of aggressive or destructive activities as well as religious and philanthropic activities.

The only escape from the tentacles of loneliness and boredom and entry in to the domain of happiness and bliss is realizing your real self, which is what you are.

The real self is boundless, timeless and is a state of bliss. Everyone has it inside oneself and the problem is we make every conscious effort to stay away from it.

Self-Image and Self

The time we are born, we are bestowed with identities we are destined to live with. This process may start with the person's name, the very first act done as soon as a baby is born. Almost always, the name itself carries the seal of the religion, gender, race, nationality, language, ethnicity and other local and regional hallmarks. Since these identities are thrust on us from infancy and maintained by powerful socio cultural norms, they exert a draconian hold on us. Later, the person's identity can be reinforced or refurbished based on his family status, social status, occupation and roles he takes.

In course of time, a person builds an image of himself with the identities he was born with as well as the credentials he has earned. It may be something of the profession like 'I am a doctor or entrepreneur or executive.' It may be a social position like 'I am social worker or home-maker or religious person etc.'

Most people live these social identities and images and not their real selves. Leaving these identities incurs a great deal of wrath of the society, which can make life miserable and impossible for the person. A society which is ordained and trained to think in a particular way and act in specific manners thinks that it is its duty to correct any such 'deviant' behavior and use all coercive methods at hand to enforce its agenda.

For the person concerned, dropping the identity one has assiduously built over time will also be a great challenge. It shatters their self-image, makes all their value systems look unworthy and can destroy their self-esteem.

The False Self is an artificial persona or image that people create very early in life to protect themselves from experiencing developmental trauma, shock and stress in close relationships. The self-image of most people is unhealthy false self, one that fits into society through forced compliance rather than a desire to adapt. Breaking this image shell and coming out needs a lot of courage and conviction. It is a process and cannot be done overnight.

Society as Source of Disintegration

Primitive societies were formed by small groups of people. They formed bands purely on personal contact and bonding, it is said the maximum manageable group size was only about 150 members. Above this size, they became unmanageable and tended to splinter out.

Survival of the fittest is an obvious phenomenon seen in all life forms. Modern humans have to struggle for their survival and also have to compete with every other person for the meager resources existing in this world. They also have an inherent predisposition to anger, aggression, possessiveness and desire to dominate. It is a common sight to see them go on very frequent ego rides to assert their power or for umpteen other reasons.

Then how did millions and billions of people live in massive groups in modern times?

People need reasons to come together. It is very difficult to pull together people on good will alone, which has very little traction on most of the people. They need additional motives for the same.

So they are given perceived identities to form larger groups. The most common of them are religions, nations, languages, races, tribes and similar entities.

Yuval Noah Harari in his famous book 'Sapiens – A brief History of Humankind' says "Large number of strangers can operate successfully by believing in common myths."

Describing the fallacy and subjective nature of believing in such entities, Harari also adds "Yet none of these things exists outside the stories that people invent and tell one another. There are no gods in the universe, no nations, no money, no human rights, no laws and no justice outside the common imagination of human beings."

But the pity is, these myths not only serve to form huge groups and cooperate with each other, they also give them reasons to fight and annihilate other groups of human beings.

More wars, violence and murders of fellow human beings have happened over religion than any other issue. Myths about one's own 'nation' and the 'patriotism' make millions of people take to arms and kill each other.

When the European powers colonized the world through the last few centuries, they made it a point to divide the indigenous people on ethnic, religious, racial, tribal and whatever lines possible. This made the colonized people busy fighting each other and prevented them from raising their voice against the occupiers. 'Divide and rule' was their policy. The divisions they created in humanity are so deep, the former colonies are still shedding blood over the issues, even though the aggressors are long gone.

Modern political parties the world over aggravate these divisions for gaining traction over the masses and make political gains.

Religious myths are so strong, it is possible to take people on a ride based on them alone. The mythical character Lord Rama has carried the mandate of about a billion people on his shoulder and landed a massive victory for a right wing political party in India.

The powerful digital media and modes of knowledge and information distribution are monopolized by the economic tycoons and political interests. They teach the people what to think and how to act. To wield power over the masses, these vested interests cause more disintegration rather than integration of humanity.

XIX. Realizing Unity

> *"Know Thyself and you will know the Universe and the Gods"*
>
> *– Inscription on the Greek temple at Delphi.*

Now, who is the person with whom you spend the maximum time with?

It is not your parents, wife, children, close friends or anybody else.

It is your Self.

If this company of yourself with your 'Self' is fulfilling and blissful, you are the most blessed person in this world. You do not need any other company at all. It does not mean you have to be alone. It just means that whether you are in a crowd, with your friends, your family or you are alone, you are at peace with yourself and enjoy every moment of your existence.

37. Know Thyself.

The Ancient Greek saying "know thyself" was inscribed in the forecourt of the Temple of Apollo at Delphi, which dates back to 4th century BC, according to the Greek writer Pausanias.

The above image is of Memento Mori mosaic from excavations in the convent of San Gregorio, Via Appia, Rome, Italy.

The Greek motto *gnōthisauton* (know thyself) combines with the image to convey the famous warning: *Respice post te; hominem teesse memento; memento mori.*

Difficulty is that modern man is a split personality. He lives one life for name, fame, material prosperity and social prestige, so that he is considered a successful person and one who has made a mark in his life. On the other hand, he has innate urges to love, to be loved, to be a wholesome and contended person. Most of the time, the first person over powers and almost destroys the basic person he is.

The first step to resolve the issue is to realize that there exists an issue. Most people do not have the patience or the time to recognize the dichotomy itself. Once this realization sets in, very likely the person will be motivated to unify his personality and realize the one unity which underlies in the depth of every person.

We go on looking outside for what we want and we go on missing it. Once we look inside, we will see that all that we had been missing so far is there. You will discover something which is changeless, timeless and eternal.

The true self is the epicenter of a person's entire being; it is the total sum of everything that we are. It typically includes the conscious and unconscious minds.

Happiness and true power lies in understanding yourself, accepting yourself, having confidence in yourself.

Your true self is deep within you, weighed down and covered by the noises of this world and ever increasing mountain of your personal garbage accumulated by you. The toughest part is to get cleared of them, and once you do it, you can discover your 'Self,' glowing like gold in your core.

Some of the simple things you can do to realize your real self can be:

Listen to Your Inner Voice. Only your heart knows what you want. Respect it and honor it.

Give Yourself the Freedom to be You, not what other people expect you to be.

Do not hesitate to make Changes in your life and explore new avenues to express your life energy. Life is a process of never ending changes. If you want life to be stable and everlasting like a rock, you can choose to be a rock rather than a being.

Have conviction on your Self and courage to follow your heart. Commit to Your Choices and Decisions and have no fear in expressing or following them.

Realizing one's Self is basically very simple. Just sit quite, be silent, relax and empty your mind of all thoughts. Your 'Self' will be there inside you to receive you with open hands. Some people call this 'Meditation.' Whatever name you give it, the principle is same. But the only problem is, dropping your junk load of thoughts and making your mind absolutely empty needs a lot of training. Though it looks difficult, it is possible and takes some perseverance.

You have to rediscover that your body is a wonder, a masterpiece of the cosmos. Your body is the seat of consciousness and it is no different from the Consciousness of the cosmos.

Mother earth is in you and father sun is also in you. You are made of sunshine, air, water, trees and mineral. Being aware of that wonder can bring you a lot of happiness.

You were born with the cosmos and will always remain an integral part of it. So your body contains all information concerning the cosmos. It also contains all the information of the history of the cosmos.

Your DNA and life source originated with the origin of life. It has been handed over through millions of species and billions of years for you to become what you are today.

In every cell of your body, you can recognize the presence of your ancestors. Not only human ancestors, but also animal, vegetation and mineral ancestors. And if you can get in touch with your body, you can get in touch with the whole cosmos – with all your ancestors and all the future generations that are already inside of your body. And that kind of awareness can be healing, can be nourishing.

This type of awareness naturally leads to unconditional love and compassion – and *unconditional love is an unlimited way of being.*

Life, through unconditional love, is a wondrous adventure that excites the very core of our being and lights our path with delight.

If we make our goal to live a life of compassion and unconditional love, then the world will indeed become a garden where all kinds of flowers can bloom and grow.

Few Things in a Nutshell

Now, let us recollect a few things we have already gone through to see whether we can come to any conclusion through this enigmatic text.

1. You are an integral part of the cosmos. The chemicals you are made of where cooked in the hearts of exploding stars.

2. You are an aggregate of attributes assembled from the cosmic dance, a phenomenon in time and space, a presence at the present moment.

3. Not only your physical body, your energy to move, talk and do all types of nonsense are

derived from the same cosmic energy. Sun's rays are the strings which move your limbs.

4. Your longing for fulfillment and wholeness is an integral, existential part of you. Your emptiness, loneliness deep inside you is universal. It awaits your acknowledgement and fulfillment.

5. The small inner voice for wholeness is drowned by the noise of this world, designed to suppress the voice and make you a bull in the yoke of the society, to carry its burden and tread the same path other bulls had been treading. The whips and yells keep you going, oblivious of your true nature, the miracle you are and the miracles around you.

6. The shackles which bind you down are the baggage you carry as concerns of the past and fears of the future. They constantly occupy your mind. Drop the baggage and move to the present. You will feel the lightness and flight. Your presence in the 'Here Now' is the only way to the realm of bliss and peace.

7. Realizing one's Self is basically simple. All it needs is dropping your junk load of thoughts and identities and making your mind absolutely receptive. Though it looks difficult, it is possible and takes some perseverance.

8. Your consciousness is part of the cosmic consciousness, which is already in you. Once you realize the barriers and gaps in between, they disappear like fog, driving you into the super conscious state.

9. Self-Realization is the key to your happiness and enlightenment.

10. The only option for your deliverance from the inevitable and terrifying human loneliness is love.

11. Love, or in its broadest sense, compassion, connects you to the people and the world at large in a meaningful way.

12. Practiced in the proper way, Love can give you an unimaginable sense of wholeness and happiness which no other achievement in your life can ever give you.

13. Science has shown in unequivocal terms the oneness of this cosmos and the unitary nature of the existence.

14. You are an integral and inseparable part of this cosmic process which goes beyond time, space and yourself.

Any deep exploration of your Self, through love and meditation will ultimately reveal that Time, Space and You are mere concepts, existing more in your mind than in reality.

Once you realize this, you enter into the timeless, boundless, formless realm of being.

This super consciousness will be your source of unlimited happiness and peace of mind for ever.

References and Further Reading

1. Karen Armstrong, "A History of God – The 4000 year quest of Judaism, Christianity and Islam," Random House Publishing Group, 1994.

2. Coleman Barks, "Rumi: The Big Red Book," Harper Collins, 2011.

3. Brian Cox, Andrew Cohen, "Human Universe," Harper Collins, 2014.

4. Charles Darwin, "Descent of Man," Second Edition, 1874.

5. Richard Dawkins and Yan Wong, "The Ancestor's Tale," Orion Books, 2010.

6. Albert Einstein, "Essays in Humanism," Philosophical Library/Open Road, 2011

7. Victor E. Frankl, "Man's Search for Meaning', Beacon Press, 1959.

8. Sigmund Freud, "The Future of an Illusion," 1928, Digital Edition, Amazon Kindle, 2013.

9. Erich Fromm, "The Art of Loving," Open Road Integrated Media, 2013

10. Erich Fromm, "Escape from Freedom," Open Road Media; Owl Book edition, 2013

11. Stephen W. Hawking, "A Brief History of Time," Bantam Books, 1989.

12. Yuval Noah Harari, "Homo Deus – A Brief History of Tomorrow," Vintage, 2017.

13. Yuval Noah Harari, "Sapiens, A Brief History of Humankind," Penguin Random House, UK, 2014

14. Stephen Hawking and Leonard Mlodinow, "The grand Design," Bantam Press, 2010.

15. Brad Jenson, "The Universe Always says Yes," Amazon Asia-Pacific Holdings Private Limited, 2015.

16. Johnstone, B., Bodling, A., Cohen, D., Christ, S. E., & Wegrzyn, A. (2012). Right parietal lobe-related "selflessness" as the neuropsychological basis of spiritual transcendence. International Journal for the Psychology of Religion. accessed at http://www.tandfonline.com on 5/30 2012.

17. Carl Gustav Jung, "Modern Man in Search of a Soul," Routledge classics, 2001.

18. Michio Kaku, "Hyperspace," Oxford University Press, 1994.

19. Ted Nield, "Supercontinent – Ten Billion Years in the Life of our Planet," Granta Publications, 2007

20. David Reich, "Who We Are and How We Got Here – Ancient DNA and the New Science of Human Past," Pantheon Books, New York.

21. Matt Ridley, "The Red Queen – Sex and the evolution of Human Nature," Harper Perennial; 2nd edition, 2012.

22. Benjamin J. Sadocket. al., Kaplan &Sadock's Comprehensive Textbook of Psychiatry, Wolters Kluwer, 2017.

23. Penny Spikins, "How Compassion Made us Humans," Pen and Swords Books Ltd., 2015

24. E.W.F. Tomlin, 'Philosophers of East and West,' Oak – Tree Books Limited, London, 1986.

Picture Credits

1. Thilo Parg – Own work, CC BY-SA 3.0, https://commons.wikimedia.org/w/index.php?curid=29686619

2. https://commons.wikimedia.org/wiki/File:GabillouSorcier.png

3. Unknown, Public Domain, https://commons.wikimedia.org/w/index.php?curid=6040879

4. Jeff Dahl – Own work, CC BY-SA 4.0, https://commons.wikimedia.org/w/index.php?curid=3257647

5. Picture of a model kept at the Chennai Museum, India – Photograph by the Author.

6. http://go.nasa.gov/2lDq44w

7. Quarks, Proton and Neutron - Original

8. What are you? – Original

9. Wheel of Time, as per Hindu Mythology- Original

10. Calabi yau formatted. – Wikimedia Commons, the free media repository

11. *Image Credit: X-ray: NASA/CXC/RIKEN & GSFC/T. Sato et al; Optical: DSS*

12. *Image Credit: NASA Formation of the Solar System: Birth of Worlds*

13. Difference DNA RNA-DE.svg: Sponk/Chemical structures of nucleobases by Roland1952, CC BY-SA 3.0, https://commons.wikimedia.org/w/index.php?curid=9810855

14. Plants growing at the edge of water – Photograph – Original

15. DNA -Public Domain, https://commons.wikimedia.org/w/index.php?curid=510095

16. Photograph – by Author

17. Rocky Mountain Laboratories, NIAID, NIH - NIAID:

    https://commons.wikimedia.org/w/index.php?curid=104228

18. *Zina Deretsky, National Science Foundation*

19. *File:Hunting Woolly Mammoth.jp – Wikipedia, the free encyclopedia*

20. *https://commons.wikimedia.org/wiki/File:Early_human_history_(book_illustration).jpg*

21. *Original*

22. *Photograph by Author*

23. *A newspaper clipping*

24. *Original*

25. *Original*

26. *Original*

27. *Original*

28. *Human Evolution Timeline: Pinterest – humanevolutionofficial.weebly.com*

29. *By User: Conquistador, User:Dbachmann – updated version of File:Homo-Stammbaum, Version Stringer-en.svg, CC BY-SA 4.0, https://commons.wikimedia.org/w/index.php?curid=64917473*

30. *By Altaileopard, Public Domain, https://commons.wikimedia.org/w/index.php?curid=34697001*

31. *By Kious, Jacquelyne et al.*

    http://pubs.usgs.gov/gip/dynamic/historical.html., Public Domain, https://commons.wikimedia.org/w/index.php?curid=6314873

32, 33. *Photos by Author, 2012*

34. *Photo by Author, 2014*

35. *Photo by Author, 2016*

36. *Photo by Author, 2020*

37. *-Lessing Photo Archive: http://www.lessing-photo.com/p3/110103/11010329.jpg, Public Domain, https://commons.wikimedia.org/w/index.php?curid=17093291*

About the Author

Dr. Azar M.B.B.S., D.A., M.D., Ph.D. is an Anaesthesiologist and a Pain and Palliative Care Physician by profession. He held teaching jobs as professor at Medical Colleges in Chennai, India for the past three decades. He is now an Honorary Professor – Consultant in Pain and Palliative Care at Cancer Institute, Adyar, Chennai, India. He has several research papers and presentations in the medical field to his credit.

Currently, he also works with Medecins Sans Frontieres (Doctors Without Borders) and has rendered humanitarian work in war zones and difficult places like Syria, Yemen, South Sudan and Papua New Guinea.

Spirituality is his passion and this work is an attempt to share his spiritual and worldly rendezvous with the world at large.

You can Contact Author By email – newrayfn@gmail.com

www.ingramcontent.com/pod-product-compliance
Lightning Source LLC
Chambersburg PA
CBHW030940180526
45163CB00002B/644